A MORPHOLOGICAL
ATLAS OF
INSECT LARVAE

A MORPHOLOGICAL ATLAS OF INSECT LARVAE

By

H. STEINMANN C. Sc. (Biol.)
and
L. ZOMBORI

AKADÉMIAI KIADÓ, BUDAPEST 1984

Manuscript revised by

Z. KASZAB

Member of the Hungarian Academy of Sciences

and

G. SZELÉNYI D. Sc. (Biol.)

Latin text revised by

L. PINTÉR

ISBN 963 05 3417 7

© Akadémiai Kiadó, Budapest 1984

Printed in Hungary

CONTENTS

Introduction 7

Plates 9

Anamorphosis 10
    Anametabolia 10
        Protura 10

Holomorphosis I. 17
    General part 18
    Special part 41
        Ametabolia 41
            Collembola 41
            Diplura 48
            Thysanura 55
        Hemimetabolia 62
            Ephemeroptera 62
            Odonata 70
            Plecoptera 81
        Paurometabolia 90
            Orthoptera 90
            Phasmoidea 100
            Dermaptera 107
            Blattaria 115
            Mantoidea 121
            Isoptera 127
            Zoraptera 135
            Embioptera 140

Parametabolia 148
    Psocoptera 148
    Mallophaga 155
    Anoplura 163
    Thysanoptera 167
    Heteroptera 173
    Homoptera 182

Holomorphosis II. 191
    General part 193
    Special part 206
    Holometabolia 206
        Coleoptera 206
        Strepsiptera 229
        Hymenoptera 234
        Megaloptera 248
        Raphidioptera 253
        Neuroptera 258
        Mecoptera 268
        Trichoptera 272
        Lepidoptera 289
        Diptera 308
        Siphonaptera 322

Indexes 327
    Latin–English 329
    English–Latin 365
    Selected references 401

INTRODUCTION

The compilation of the atlas illustrating the general external morphology of the fully developed insect was the outcome of a natural demand arriving from the field of identification researches. After the publication of a homologized atlas of the adult insect forms (Steinmann and Zombori, 1981) it seemed inevitable to undertake the task of elucidating the general terminology of the postembryonic forms, too. As we had done in our previous work, the structure of the present atlas follows suit: for easy handling and quick recognition purposes a simple structure has been adopted. Using a wide selection of the literature, the authors have endeavoured to gather all relevant material in order to introduce students of morphology to the realm of the external characteristics of insect larvae.

The material of the entire atlas is fully homologized: the treatment is based on both ontogenetics and systematics. The divisions of metamorphosis, with slight modifications, follow those of Beier (1969) and Fischer (1969). Thus, from the anamorphosis and the holomorphosis main groups we could draw up the phylogenetically and systematically accepted order of Insecta.

The holomorphosis main group has been subdivided into two units, each introduced by a general part, to show the common characteristics of the larvae and nymphs, respectively. Many a figure has been designed newly, mostly from dry, partly from alcoholic material. When choosing a specimen for illustration we consistently selected a specimen of an advanced stage of development.

The morphological treatment of the larvae, besides the two homologized parts, follows a systematic sequence of orders giving line drawings numbered consecutively. The thus shown material to some degree tries to loosen up the very strictly homologized general parts with many schematic figures. Within the insect orders we declined to use specific names, and the lowest category, usually representing a type, has been the genus. In an order we selected only those representative figures which might best illustrate the characteristics common to the principal types.

In order to identify any particular part of the body with certainty the terminology is given in Latin. In including a term into the atlas our governing principle was to choose the most descriptive word or expression whose functional content is readily conceivable among the host of synonyms widely used in the technical literature throughout the world. At this point we should like to stress that when a name was selected to describe a bodily feature, no overruling decision had been made for the supremacy of that particular term, in other words, the authors had no intention of suppressing any one word or expression. It is all the more important to bear this in mind since many specialists, including all the applied branches of related sciences working on a special group of insect larvae, adopted a terminology of their own and these terms have a standing of several decades or perhaps of two centuries. Consequently, several synonymous names and expressions have been entered into the figures, and may be found side by side, of course, in the indexes, since in that special group of larvae, usually covering a clearly delimitable unit, that name is in current use, and not the other equally valid parallel expression.

The atlas part proper is followed by two indexes: a Latin–English and an English–Latin one. The indexes are arranged in an alphabetical order, and the Latin words are given as they appear in the plates. After the word-pairs the numbers refer to the number of the figures.

The book is closed by a very limited number of literature references. Since the morphological literature of insect larvae is wide, and especially the number of papers is rather high, we decided against referring to all of those which we had consulted. However, a selected list of references is given to those names which appear at the end of each figure caption. When a figure has been borrowed from an earlier work, we refer only to our source.

PLATES

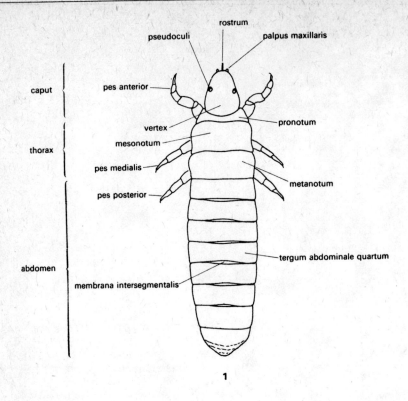

Fig. 1. Schematic drawing of the Protura in dorsal view

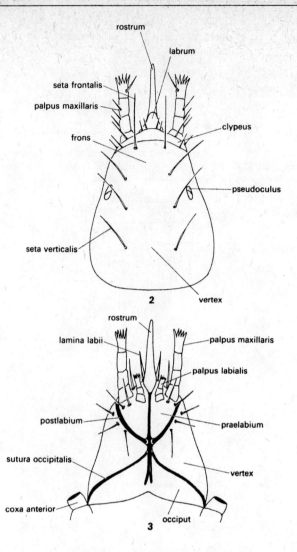

Figs. 2–3. Head of Protura in dorsal and ventral view (after Janetschek, modified)

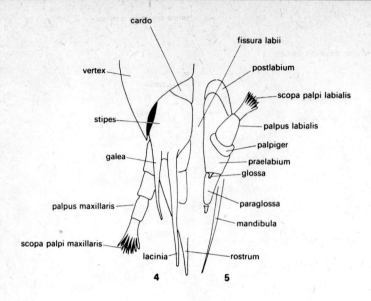

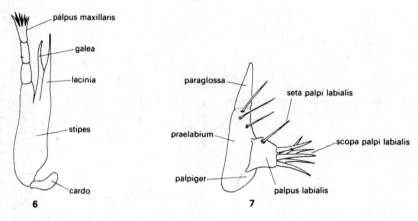

Figs. 4–7. Mouthparts of Protura in ventral view. – 4–5: Maxilla and labium intact. 6: Maxilla of the Proturentomon-type. 7: Labium of the Acerentulus-type (after Janetschek, modified)

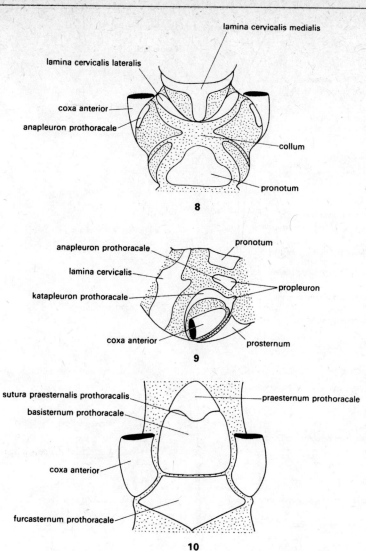

Figs. 8–10. Prothorax of Protura in dorsal, lateral and ventral view (after Janetschek, modified)

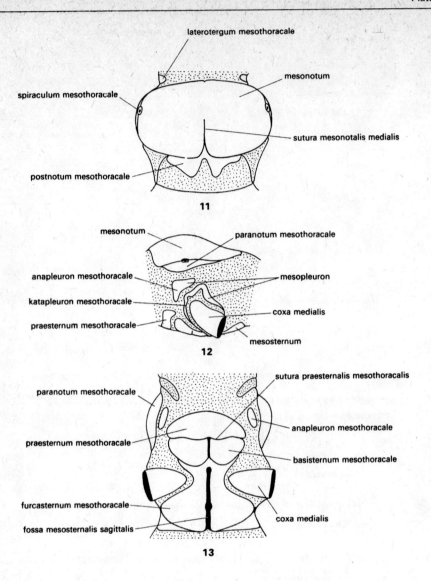

Figs. 11–13. Mesothorax of Protura in dorsal, lateral and ventral view (after Janetschek, modified)

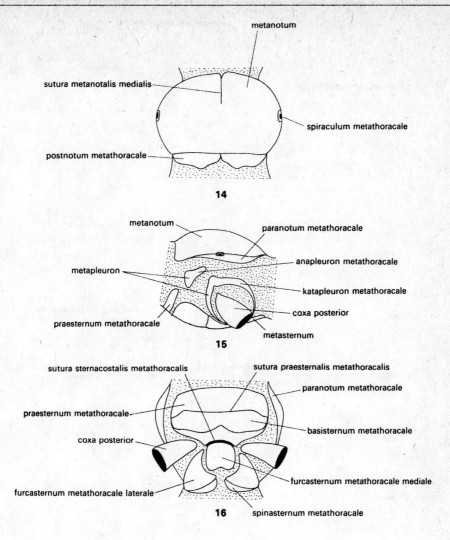

Figs. 14–16. Metathorax of Protura in dorsal, lateral and ventral view (after Janetschek, modified)

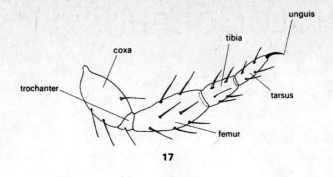

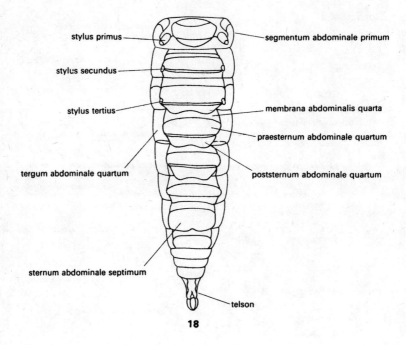

Fig. 17. Typical thoracic leg of Protura. – Fig. 18. Abdomen in ventral view (after Janetschek, modified)

HOLOMORPHOSIS I.

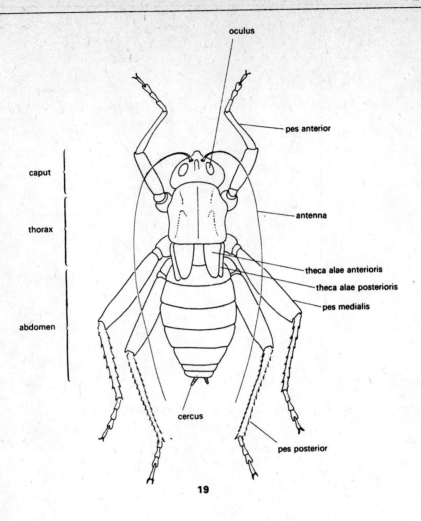

Fig. 19. Primary nymph in dorsal view

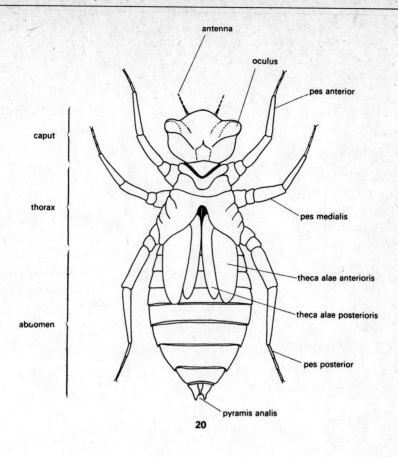

Fig. 20. Secondary nymph in dorsal view

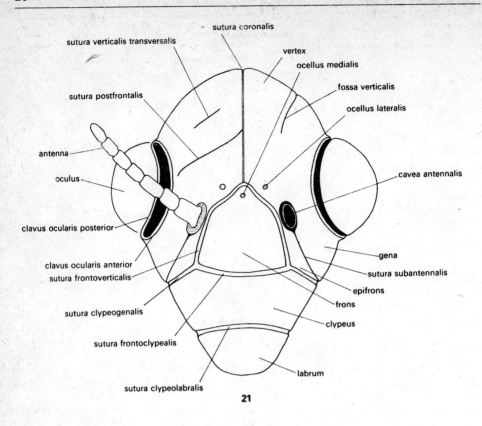

Fig. 21. Schematic drawing of the nymph head in frontal view

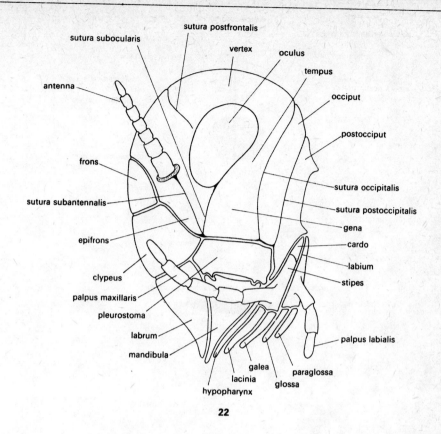

Fig. 22. Schematic drawing of the nymph head in lateral view

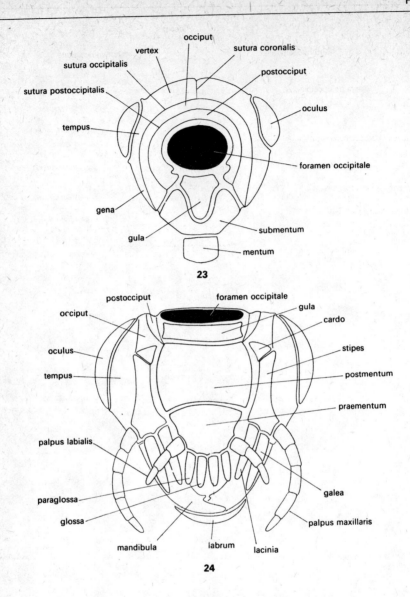

Figs. 23–24. Schematic drawing of the nymph head in posterior and ventral view

Holomorphosis I: General part

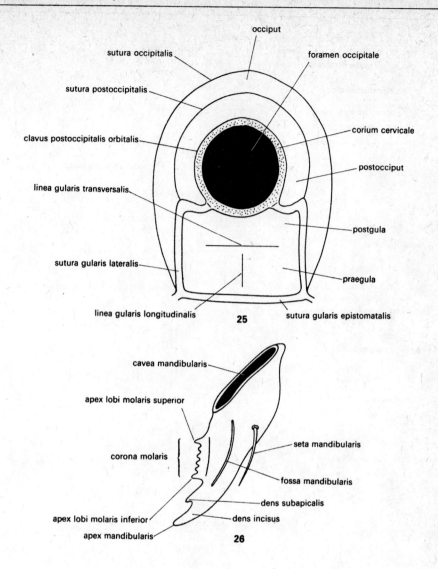

Fig. 25. The occipital and gular region (regio occipitalis, r. gularis). – Fig. 26. The mandible

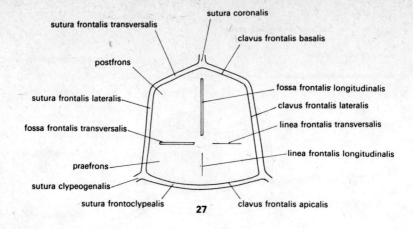
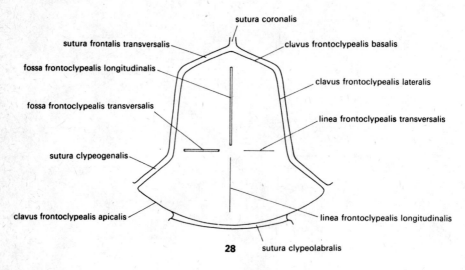

Figs. 27–28. Schematic drawing of 27: the frons, and 28: the frontoclypeus

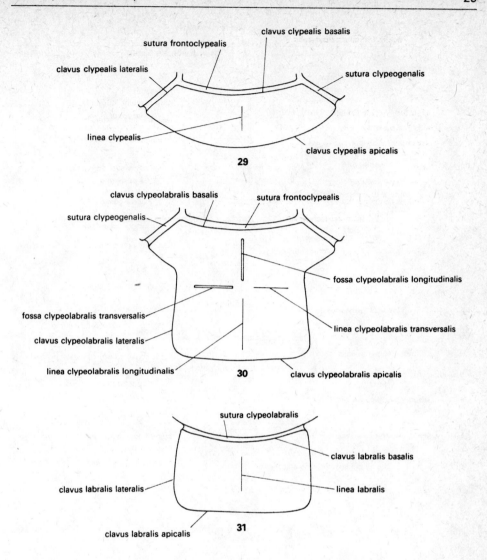

Figs. 29–31. Schematic drawing of 29: the clypeus, 30: the clypeolabrum, and 31: the labrum

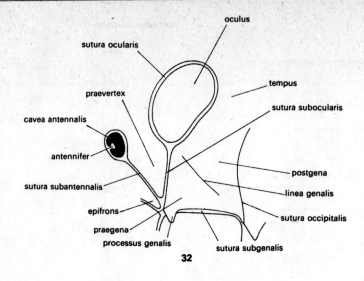

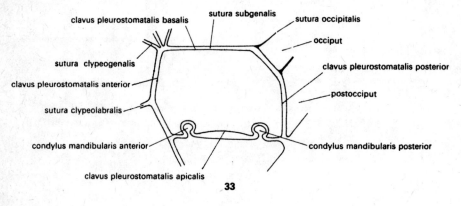

Fig. 32. The genal region (regio genalis). – Fig. 33. The pleurostoma

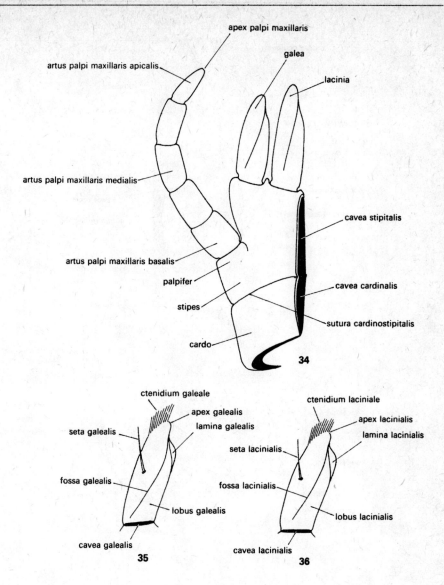

Fig. 34. The maxilla. – Fig. 35. The galea. – Fig. 36. The lacinia

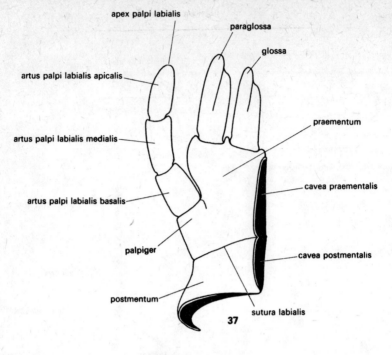
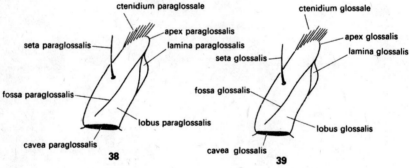

Fig. 37. The labium. – Fig. 38. The paraglossa. – Fig. 39. The glossa

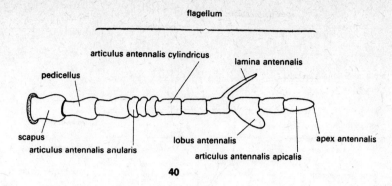

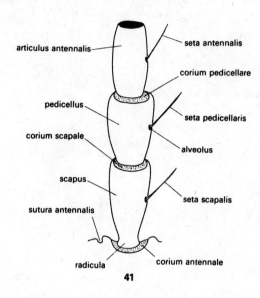

Fig. 40. Schematic drawing of the antenna. – Fig. 41. The first three antennal joints

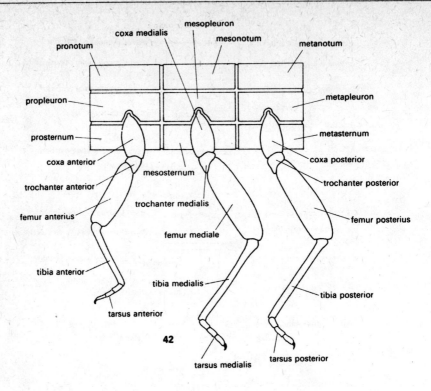

Fig. 42. Diagrammatic representation of the thorax with legs

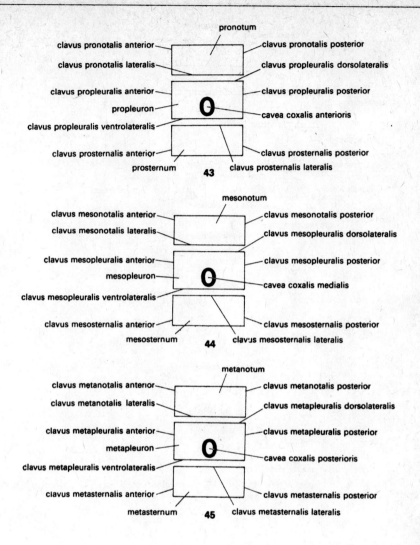

Figs. 43–45. Diagrammatic representation of the thoracic segments in lateral view

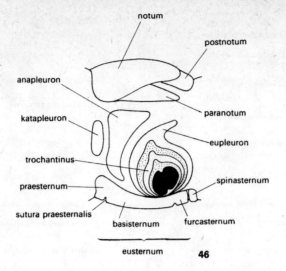

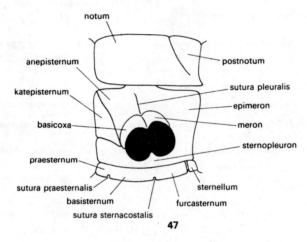

Figs. 46–47. Typical thoracic segment (segmentum thoracale) in lateral view. – 46: The Apterygota-type, and 47: the Pterygota-type

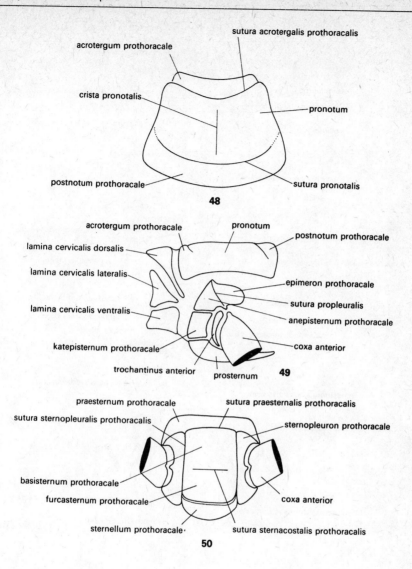

Figs. 48–50. Schematic drawing of the prothorax in dorsal, lateral and ventral view

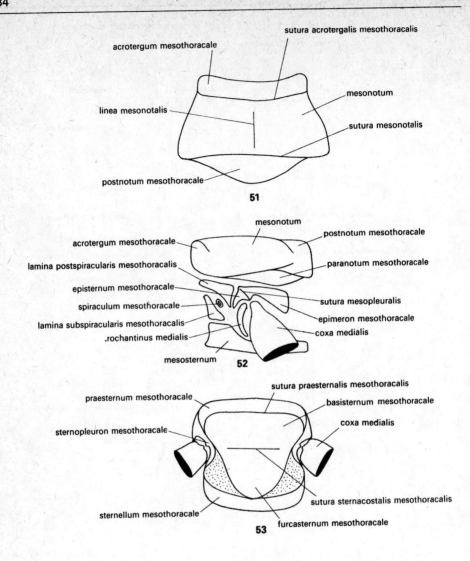

Figs. 51–53. Schematic drawing of the mesothorax in dorsal, lateral and ventral view

Holomorphosis I: General part

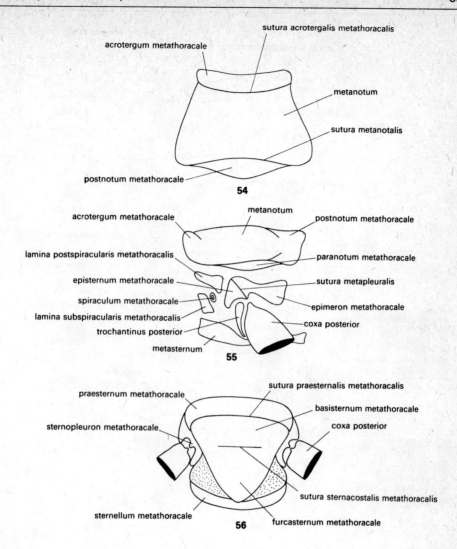

Figs. 54–56. Schematic drawing of the metathorax in dorsal, lateral and ventral view

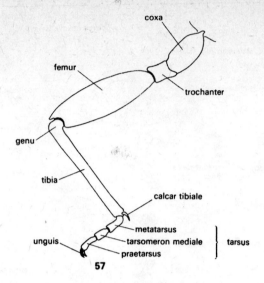
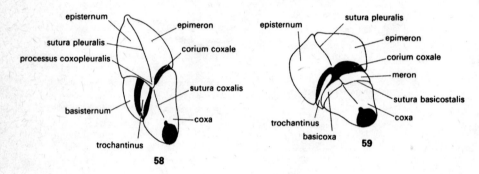

Fig. 57. Typical thoracic leg (pes). – Figs. 58–59. The coxal articulation (articulatio coxalis). – 58: articulation restricted by the coxopleural process, and 59: articulation without coxopleural process (after Snodgrass 58–59)

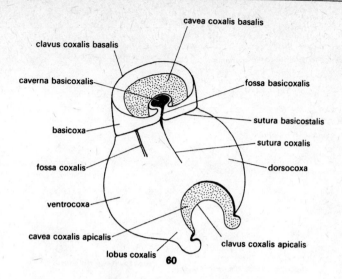
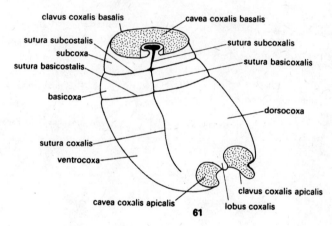

Figs. 60–61. The coxa. – 60: fixed coxa (coxa fixa), and 61: free coxa (coxa libera) (after Steinmann and Zombori)

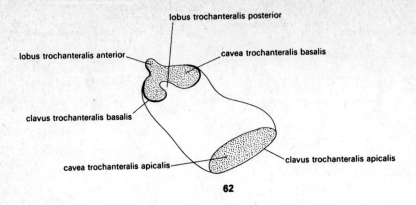

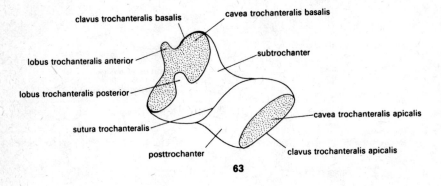

Figs. 62–63. The trochanter. – 62: simple trochanter (trochanter simplex), and 63: subtrochanter and posttrochanter (trochanter compositus) (after Steinmann and Zombori)

Holomorphosis I: General part

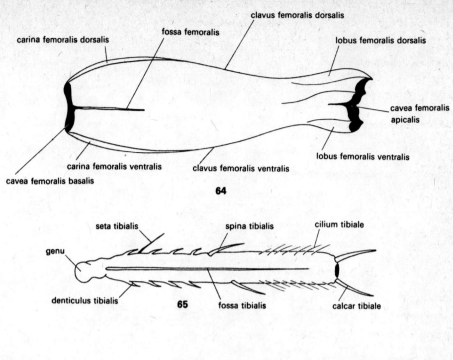

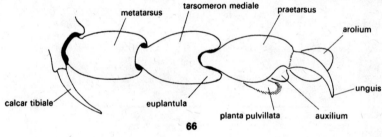

Fig. 64. The femur. – Fig. 65. The tibia. – Fig. 66. The tarsus

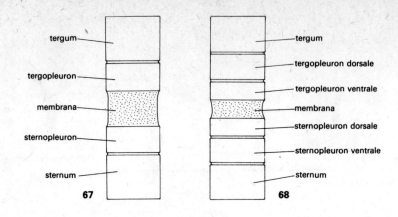

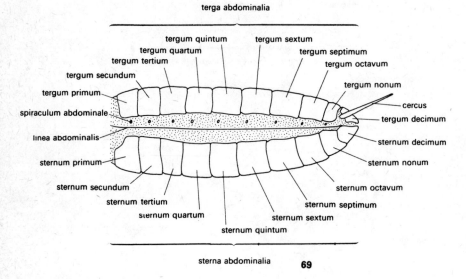

Figs. 67–68. Diagrammatic representation of the abdominal segment (segmentum abdominale). – 67: Simple, and 68: differentiated. – Fig. 69. Schematic drawing of the abdomen in lateral view (after Steinmann and Zombori 69)

Holomorphosis I: Ametabolia

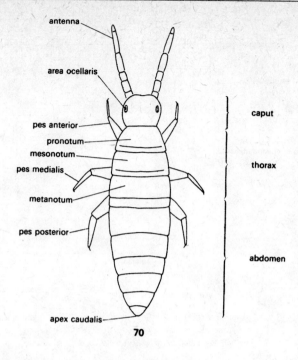

Fig. 70. Schematic drawing of the Collembola in dorsal view

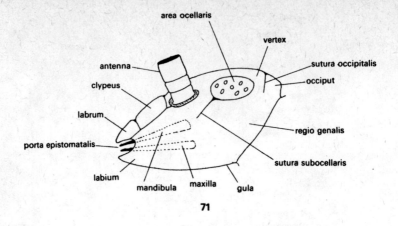

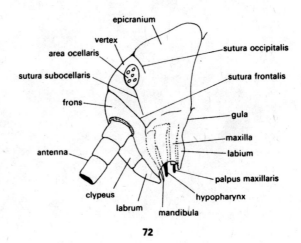

Figs. 71–72. Head of Collembola in lateral view. – 71: Prognath-type, and 72: orthognath-type (after Grassé, modified)

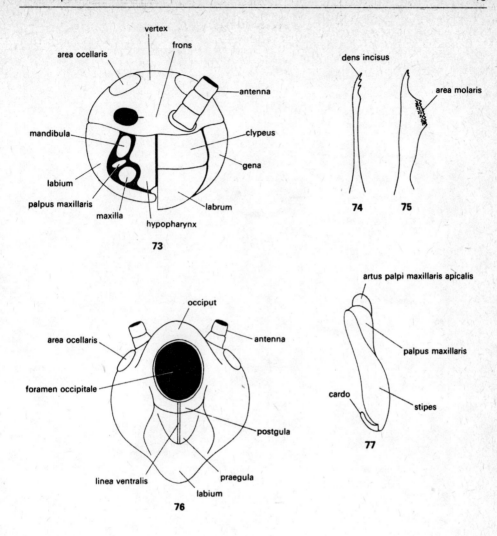

Fig. 73. Schematic drawing of the head of Collembola in frontal view (clypeus and labrum partly removed). – Figs. 74–75. The mandible. – 74: Simple, and 75: same with molar surface. – Fig. 76. Head in posterior view. – Fig. 77. The maxilla (after Schaller, strongly modified)

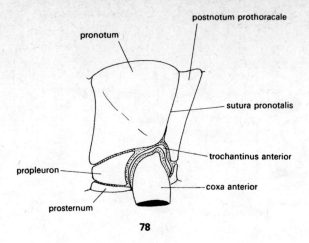

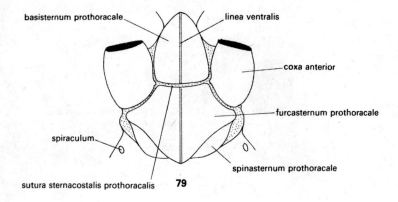

Figs. 78–79. Prothorax of Collembola in lateral and ventral view (after Matsuda, modified 79)

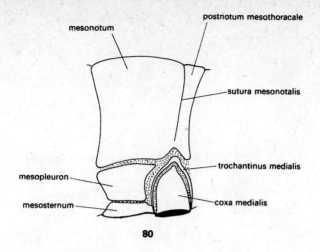

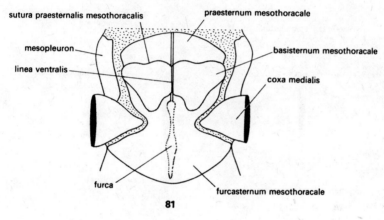

Figs. 80-81. Mesothorax of Collembola in lateral and ventral view (after Matsuda, modified 81)

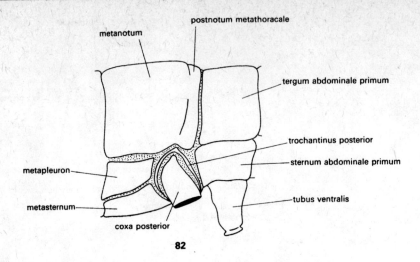

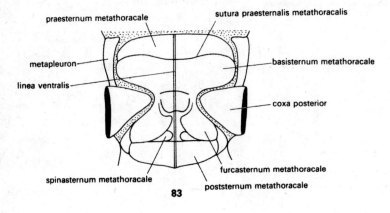

Fig. 82. Metathorax and the first abdominal segment of Collembola in lateral view. – Fig. 83. Metathorax of Collembola in ventral view (after Matsuda, modified 83)

Holomorphosis I: Ametabolia

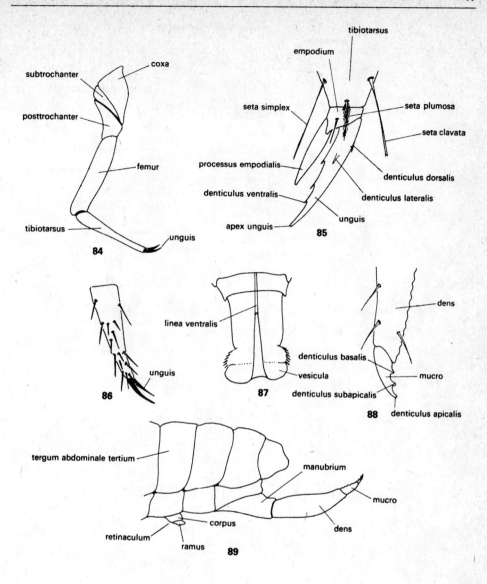

Figs. 84–89. Body appendages of Collembola. – 84: Thoracic leg, 85: the claw, 86: the claw with tibiotarsus, 87: the ventral tube in frontal view, 88: the dens and mucro, and 89: the abdominal end in lateral view (after Grassé)

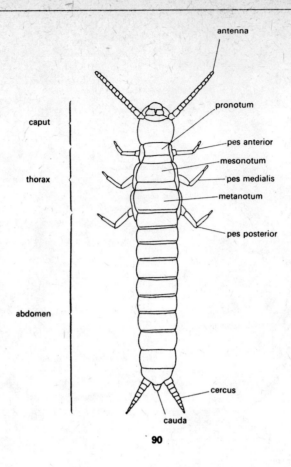

Fig. 90. Schematic drawing of the Diplura in dorsal view

Holomorphosis I: Ametabolia

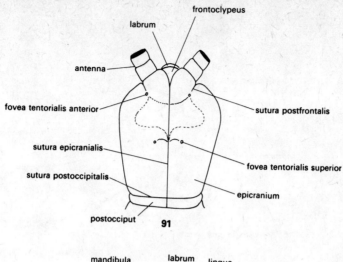

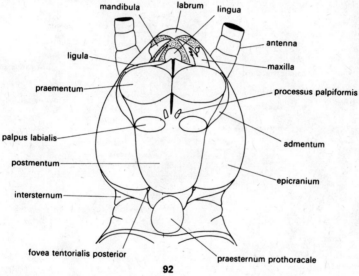

Figs. 91–92. Head of Diplura. – 91: Japyx-type in dorsal view, and 92: Campodea-type in ventral view (after Grassé 91, and Francois, modified 92)

4 Morphological

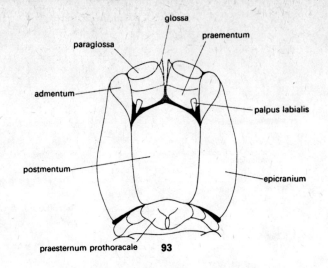
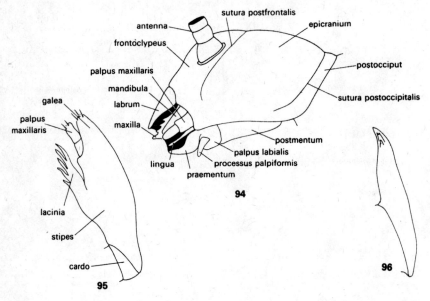

Figs. 93–94. Head of Diplura. – 93: Japyx-type in ventral view, and 94: Campodea-type in lateral view. – Fig. 95. The maxilla. – Fig. 96. The mandible (after Matsuda 93, Grassé 94, and Francois 95–96)

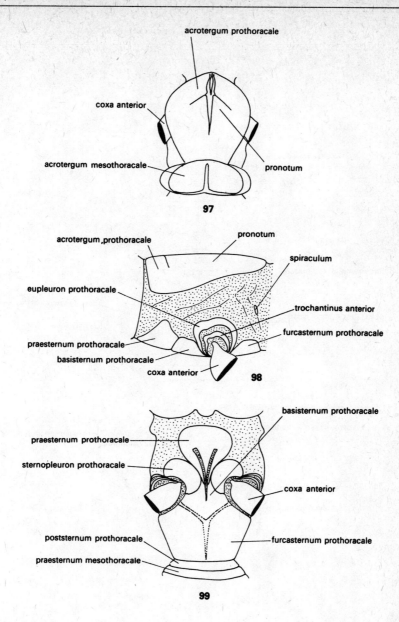

Figs. 97–99. Prothorax of Diplura in dorsal, lateral and ventral view (after Grassé, and Matsuda, modified)

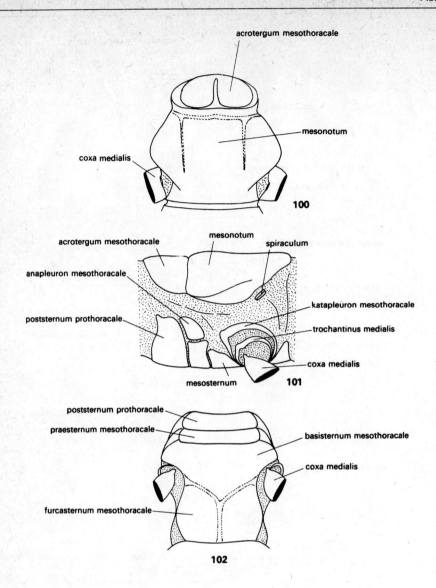

Figs. 100–102. Mesothorax of Diplura in dorsal, lateral and ventral view (after Grassé, and Matsuda, modified)

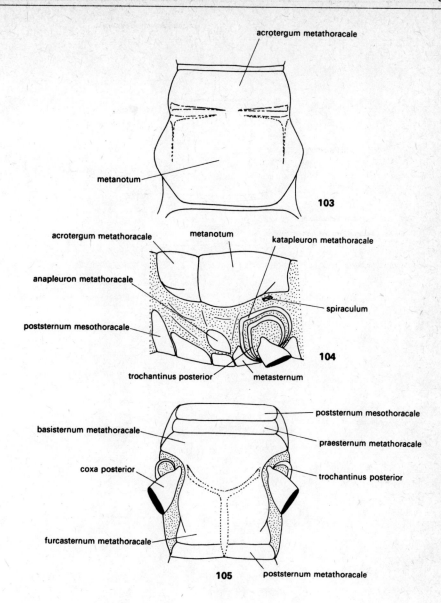

Figs. 103–105. Metathorax of Diplura in dorsal, lateral and ventral view (after Grassé, and Matsuda, modified)

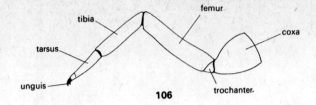

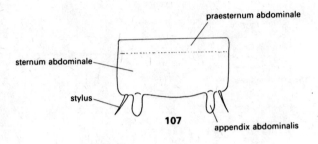

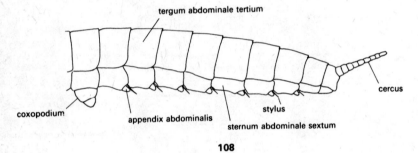

Fig. 106. The thoracic leg of Diplura. – Fig. 107. Typical abdominal sternite. – Fig. 108. The abdomen in lateral view (after Grassé 107)

Holomorphosis I: Ametabolia

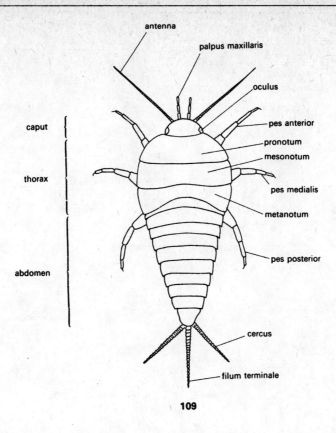

Fig. 109. Schematic drawing of the Thysanura in dorsal view

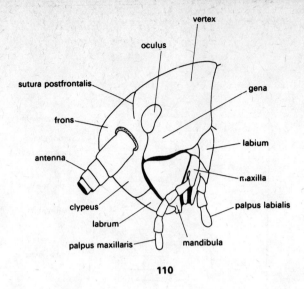
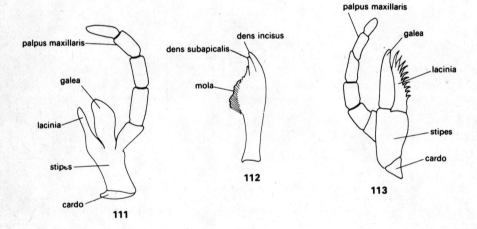

Fig. 110. Head of Thysanura in lateral view. – Figs. 111–113. The mouthparts. – 111: The maxilla of the Machilis-type, 112: the mandible, and 113: the maxilla of the Lepisma-type (after Grassé)

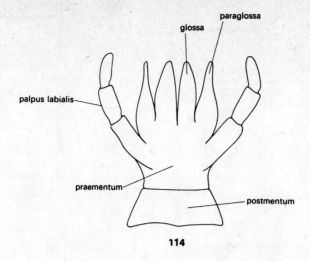

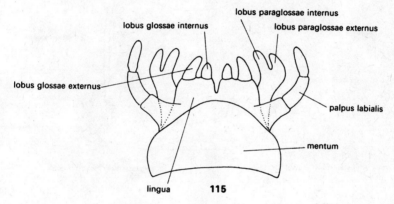

Figs. 114–115. The labium of Thysanura. – 114: Lepisma-type, and 115: Machilis-type (after Grassé)

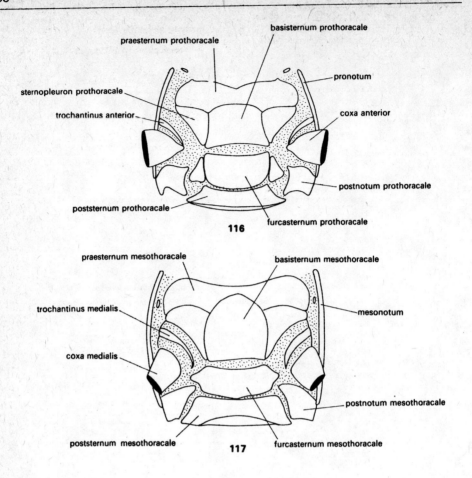

Figs. 116–117. Prothorax and mesothorax of Thysanura in ventral view (after Matsuda, strongly modified)

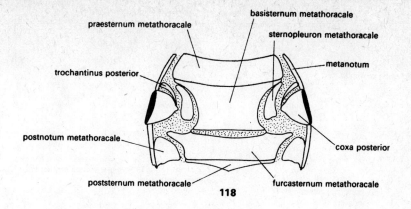

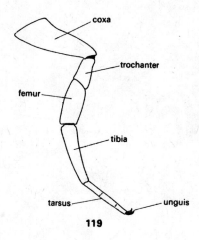

Fig. 118. Metathorax of Thysanura in ventral view. — Fig.119. The leg of the Machilis-type (after Matsuda, strongly modified 118)

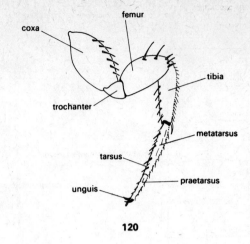

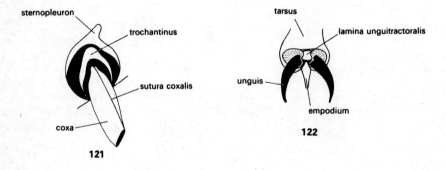

Fig. 120. The leg of the Lepisma-type. – Fig. 121. The coxal articulation of Thysanura. – Fig. 122. The tarsal end with claw

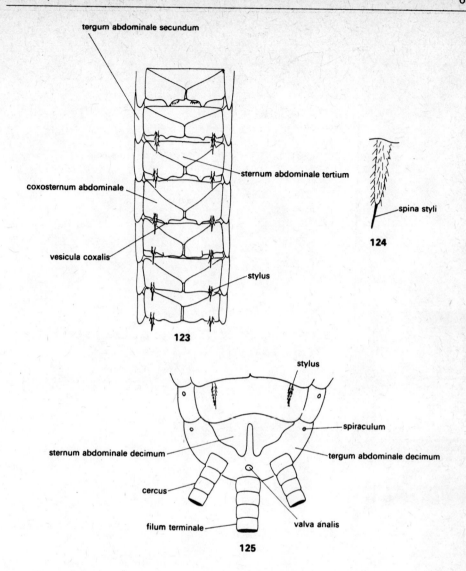

Figs. 123–125. Abdominal parts of Thysanura. – 123: Typical abdominal segments in ventral view, 124: the stylus, and 125: the abdominal end in ventral view (after Grassé)

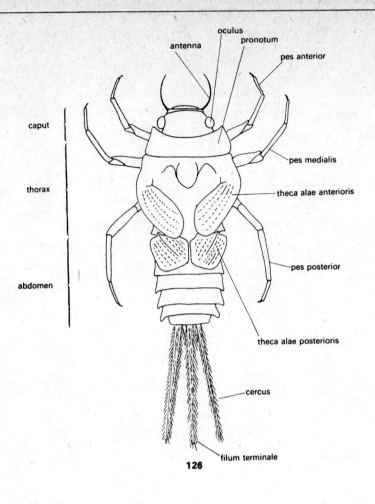

Fig. 126. Schematic drawing of the Ephemeroptera in dorsal view

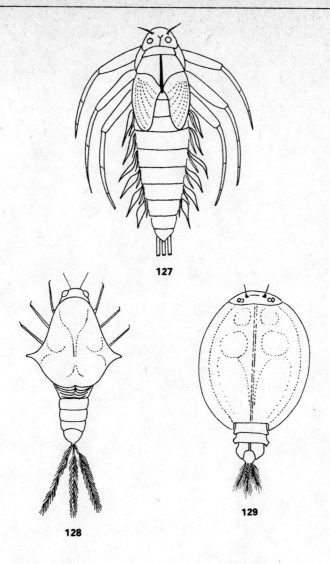

Figs. 127–129. Characteristic nymphs of Ephemeroptera. – 127: Pseudiron-type, 128: Baetisca-type, and 129: Prosopistoma-type (after Illies)

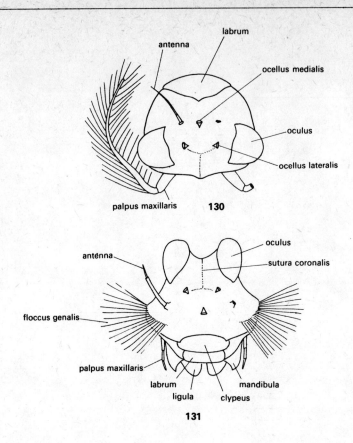

Figs. 130–131. Head of Ephemeroptera. – 130: Arthroplea-type in dorsal view, and 131: Machadorythus-type in frontal view (after Illies)

Holomorphosis I: Hemimetabolia

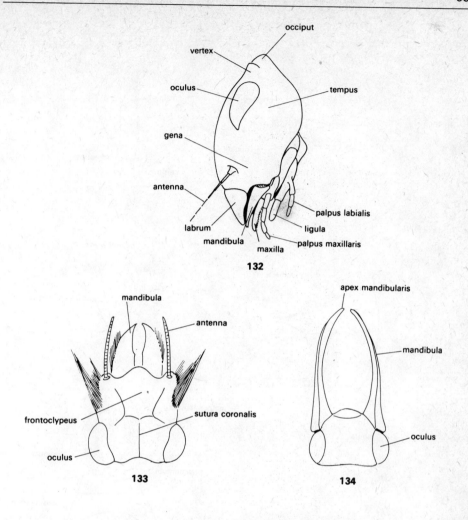

Figs. 132–134. Head of Ephemeroptera. – 132: Mirawara-type in lateral view, 133: Campsurus-type in dorsal view, and 134: Campylocia-type in dorsal view (after Illies)

5 Morphological

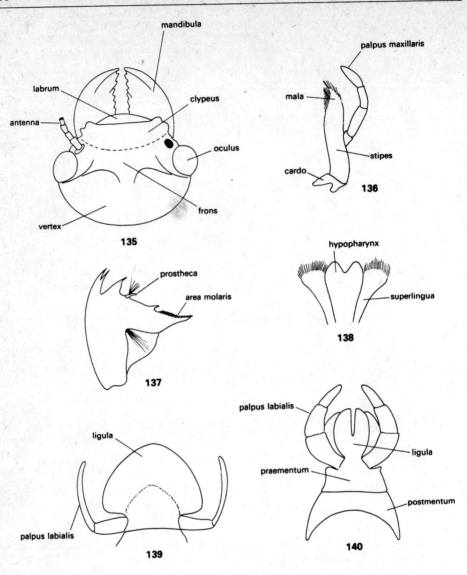

Figs. 135–140. Head and mouthparts of Ephemeroptera. – 135: Asthenopus-type in dorsal view, 136: the maxilla, 137: the mandible, 138: the hypopharynx, 139: the labium of Tricorythus-type, and 140: typical labium (after Illies 135, 139, and Grassé 136—138, 140)

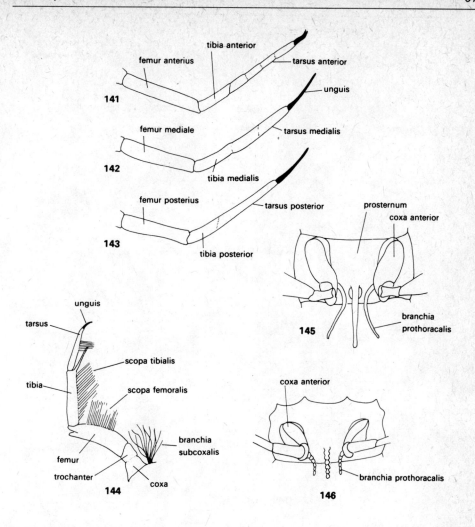

Figs. 141–143. Typical thoracic leg of Ephemeroptera. – 141: Fore leg, 142: middle leg, and 143: hind leg. – Fig. 144. Thoracic leg of the Isonychia-type. – Figs. 145–146. Prosternum of the Coloburiscus-type. – 145: Nymphal, and 146: subimaginal form (after Illies 144, and Stys and Soldan 145–146)

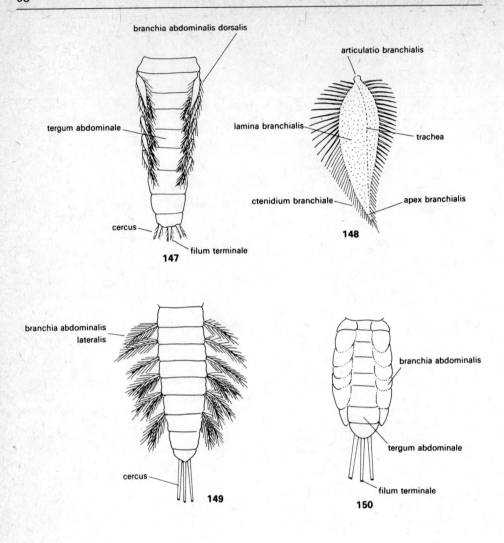

Figs. 147–150. Abdominal tracheal gills of Ephemeroptera. – 147: Abdomen of Hexagenia in dorsal view, 148: tracheal gill of Hexagenia-type, 149: abdomen of Potamanthus in dorsal view, and 150: abdomen of Ephemerella in dorsal view (after Peterson 147–148, and Bertrand 149–150)

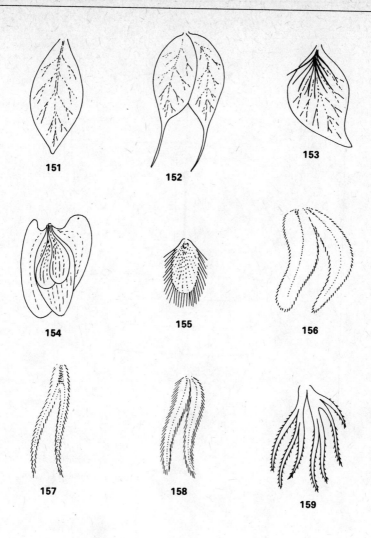

Figs. 151–159. The principal types of tracheal gills of Ephemeroptera. – 151: Choroterpes, 152: Leptophlebia, 153: Heptagenia, 154: Chitonophora, 155: Torleya, 156: Polymitarcys, 157: Habroleptoides, 158: Ephemera, and 159: Rhithrogena (after Bertrand)

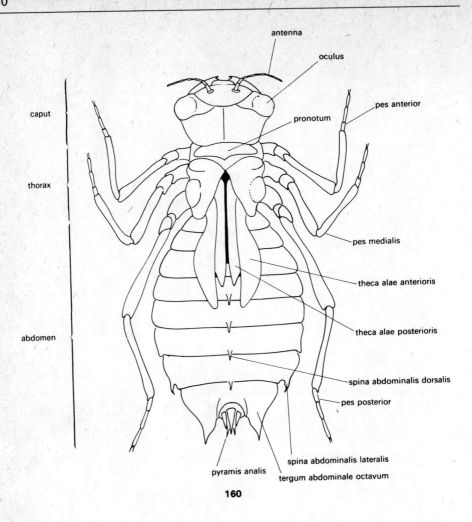

Fig. 160. Schematic drawing of the Odonata (Anisoptera-type) in dorsal view (after Peterson)

Holomorphosis I: Hemimetabolia

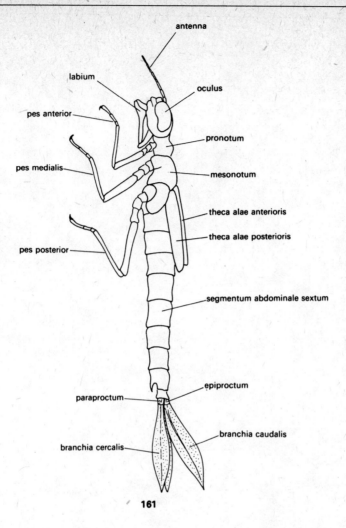

Fig. 161. Schematic drawing of the Odonata (Zygoptera-type) in lateral view (after Peterson)

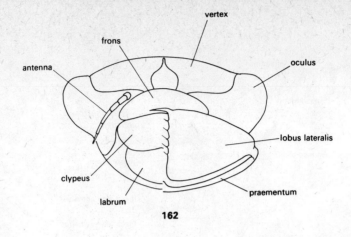
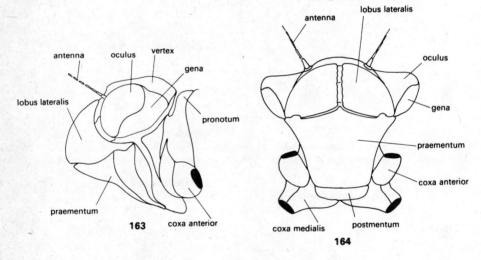

Figs. 162–164. Head of Odonata. – 162: Sympetrum-type in frontal view (labium partly removed), 163: same in lateral, and 164: in ventral view

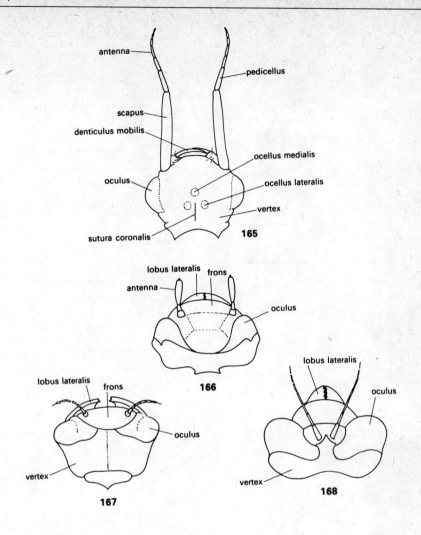

Figs. 165–168. Head of Odonata in dorsal view. – 165: Lestes-type, 166: Ophiogomphus-type, 167: Epicordulia-type, and 168: Aeshna-type (after Peterson 165–167, and Steinmann 168)

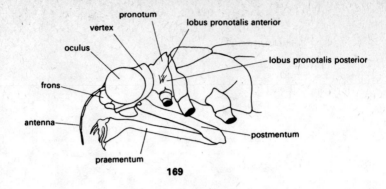

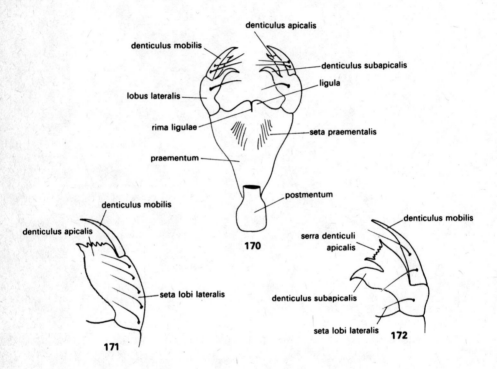

Fig. 169. Head and thorax of Odonata in lateral view. – Figs. 170–172. Parts of labium. – 170: Zygoptera-type, viewed from inside, 171: lateral lobe of Ischnura-type, and 172: same of Sympecma-type (after Peterson 169, and Steinmann 170–172)

Holomorphosis I: Hemimetabolia

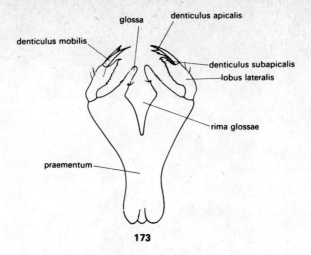

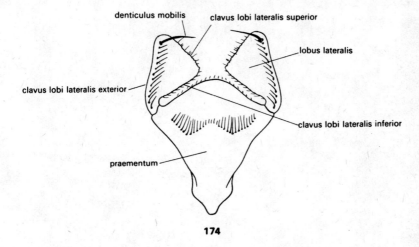

Figs. 173–174. Labium of Odonata. – 173: Calopteryx-type, and 174: Sympetrum-type (after Steinmann)

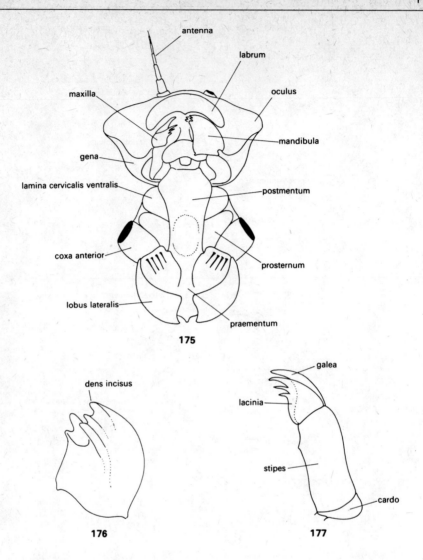

Fig. 175. Head and prothorax of Odonata (with reflexed labium). – Fig. 176: The mandible. – Fig. 177: The maxilla

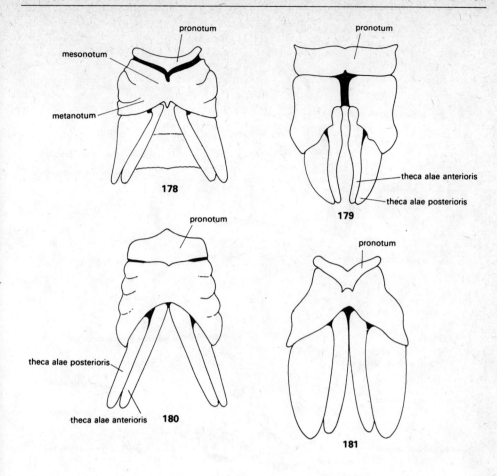

Figs. 178–181. Thorax of Odonata with wing pads in dorsal view. – 178: Cordulegaster. 179: Cordulia, 180: Erythromma, and 181: Libellula (after Steinmann)

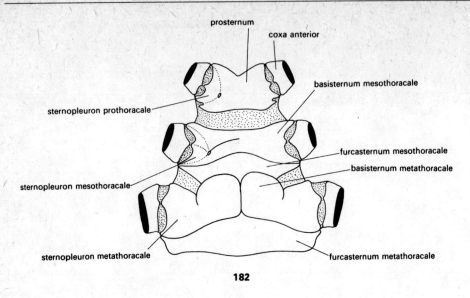

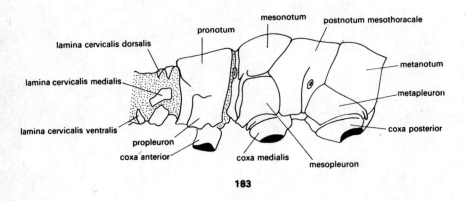

Figs. 182–183. Thorax of Odonata in ventral and lateral view (after Matsuda, modified)

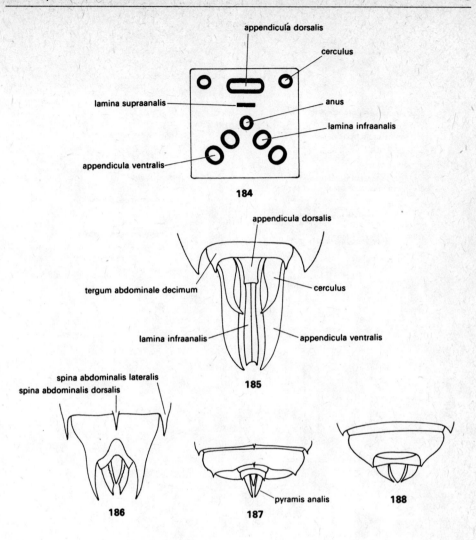

Fig. 184. Diagrammatic representation of the caudal end of Odonata (Anisoptera-type) in cross-section. – Figs. 185–188. Some types of caudal end in dorsal view. – 185: Anax, 186: Sympetrum, 187: Somatochlora, and 188: Gomphus (after Bertrand 184, Grassé 185, and Steinmann 186–188)

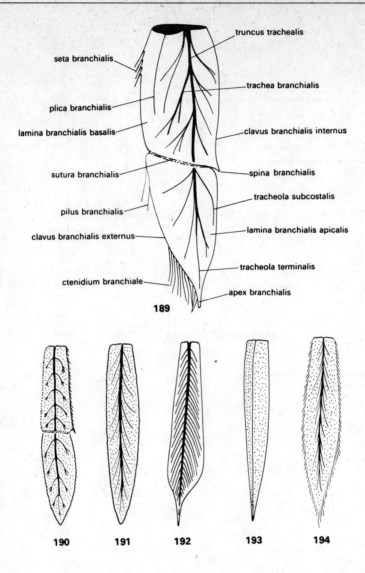

Figs. 189–194. Tracheal gill of Odonata. – 189: Schematically in detail, 190–194: some types, 190: Erythromma, 191: Ischnura, 192: Platycnemis, 193: Calopteryx, and 194: Agrion (after Steinmann 190–194)

Holomorphosis I: Hemimetabolia

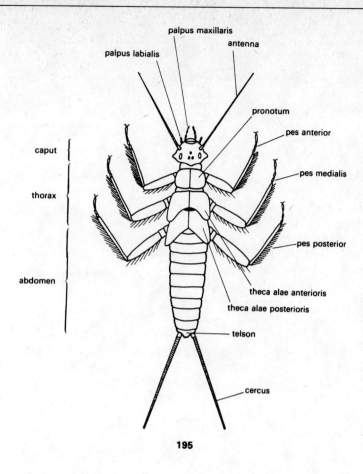

Fig. 195. Schematic drawing of the Plecoptera in dorsal view (after Steinmann)

6 Morphological

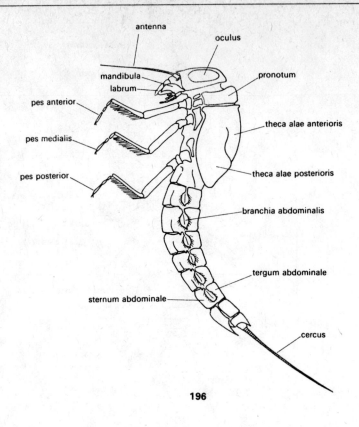

Fig. 196. Schematic drawing of the Plecoptera in lateral view (after Peterson)

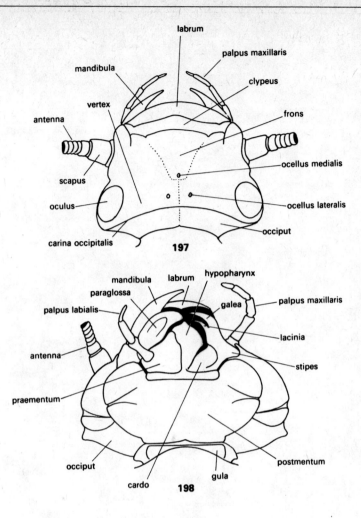

Figs. 197–198. Head of Plecoptera in dorsal and ventral view

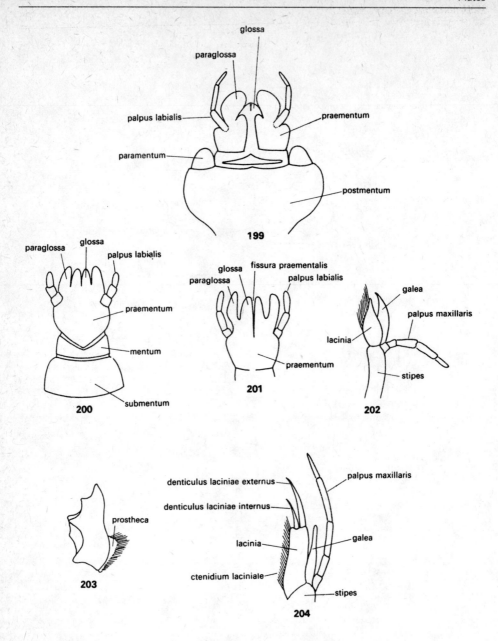

Figs. 199–204. Mouthparts of Plecoptera. – 199–201: Some types of labium, 199: Perla, 200: Capnia, and 201: Capniella. – 202: Maxilla of the Capniella-type. – 203: The mandible. – 204: Maxilla of the Arcynopteryx-type (after Steinmann)

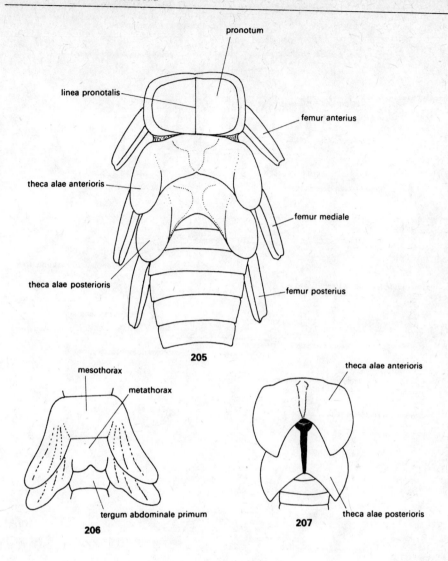

Figs. 205–207. Thorax of Plecoptera with wing pads in dorsal view. – 205: Perla, 206: Capnia, and 207: Perlesta (after Steinmann)

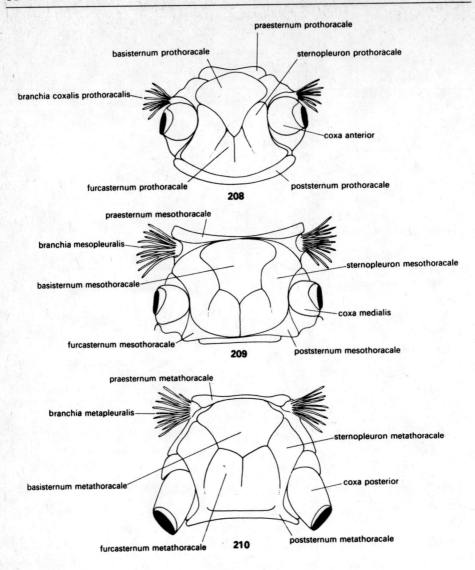

Figs. 208–210. Pro-, meso- and metathorax of Plecoptera in ventral view

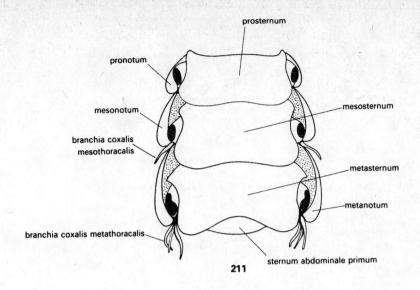

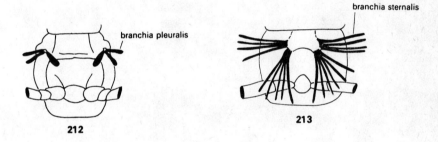

Figs. 211–213. The thorax and thoracic parts of Plecoptera in ventral view. – 211: Peltoperla, 212: Amphinemura, and 213: Nemura (after Grassé 211, Klapalek 212, and Bertrand 213)

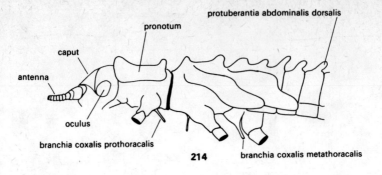

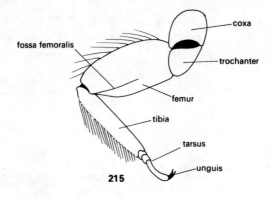

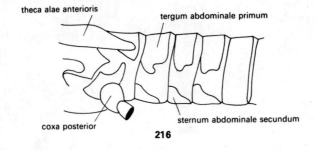

Fig. 214. Young nymph of Plecoptera in lateral view. – Fig. 215. Typical thoracic leg. – Fig. 216. Part of thorax and abdomen in lateral view (after Steinmann)

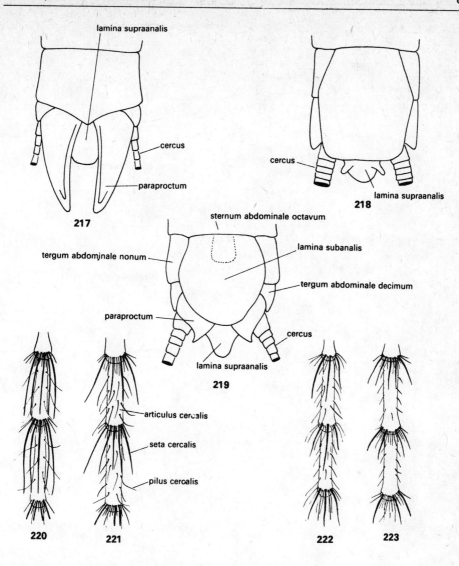

Figs. 217–219. Abdominal end of Plecoptera in ventral view. – 217: Nemurella, 218: Rhabdiopteryx, and 219: Brachyptera. – Figs. 220–223. The median cercal joints of some Amphinemura species. – 220: A. borealis Morton, 221: A. standfussi Ris, 222: A. sulcicollis Stephens, and 223: A. triangularis Ris (after Steinmann)

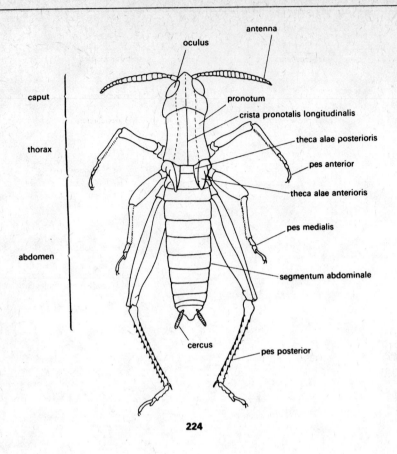

Fig. 224. Schematic drawing of the Orthoptera in dorsal view

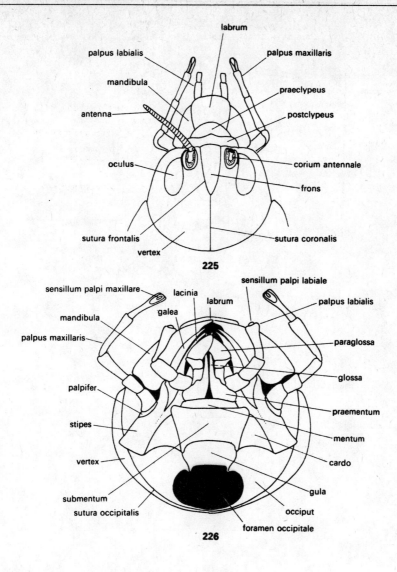

Figs. 225–226. Head of Orthoptera. – 225: Gryllus-type in dorsal, and 226: Gryllotalpa-type in ventral view

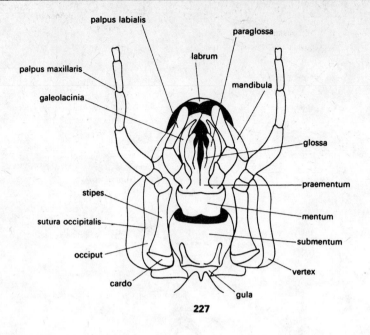

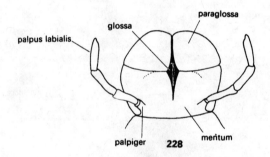

Fig. 227. Head of Orthoptera in ventral view. – Fig. 228. The labium of Schistocerca (after Beier 228)

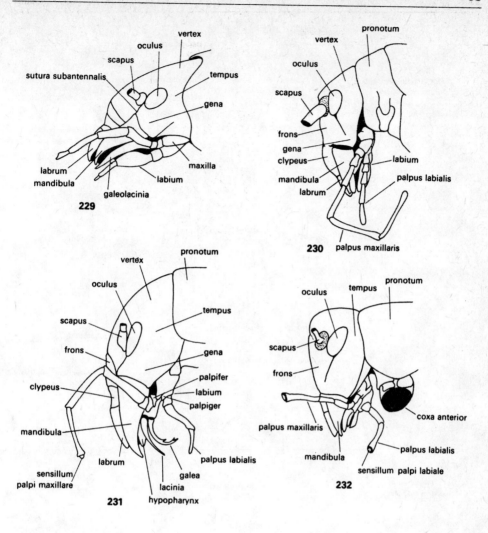

Figs. 229–232. Characteristic types of Orthoptera head in lateral view. – 229: Gryllotalpa, 230: Tachycines, 231: Brachytrypes, and 232: Gampsocleis

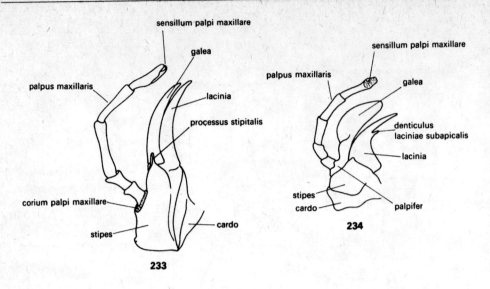
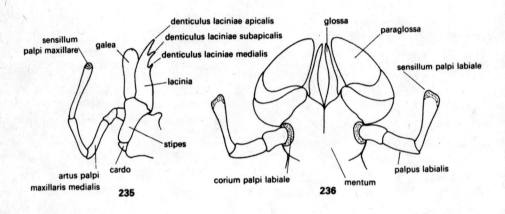

Figs. 233–236. Mouthparts of Orthoptera. – 233–235. The maxilla, 233: Gryllus-type, 234: Schistocerca-type, and 235: Tettigonia-type. – 236: The labium of Tettigonia-type (after Beier)

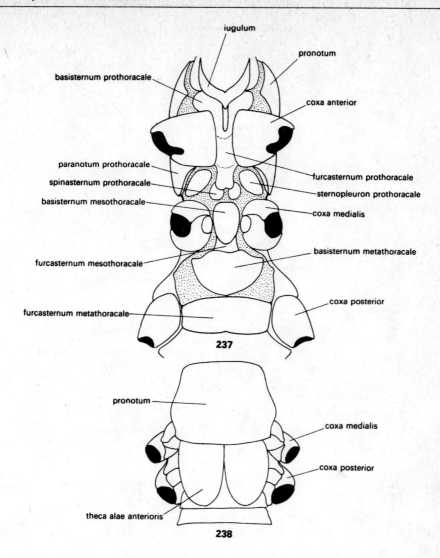

Figs. 237–238. Thorax of Orthoptera (Gryllotalpa-type) in ventral and dorsal view

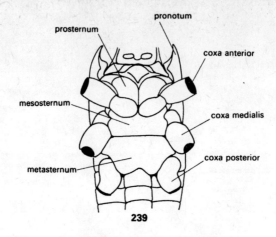

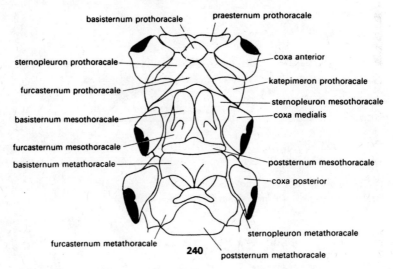

Figs. 239–240. Thorax of Orthoptera in ventral view. – 239: Gryllus-type, and 240: Gampsocleis-type

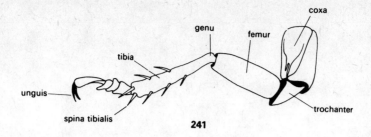
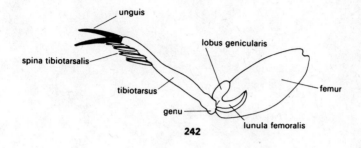
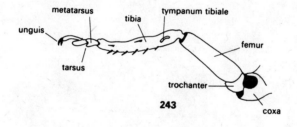

Figs. 241–243. Typical thoracic leg of Orthoptera. – 241: Brachytrypes, 242: Tridactylus, and 243: Gampsocleis (after Beier 242)

7 Morphological

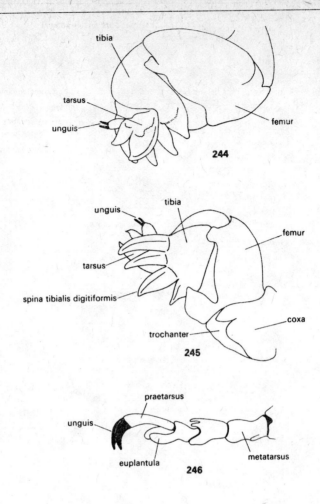

Figs. 244–245. The highly modified leg of Orthoptera: Gryllotalpa from outside and inside. –Fig. 246. The tarsus of Tettigonia (after Beier 246)

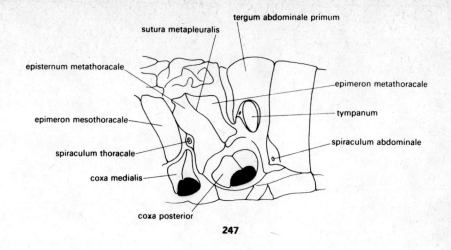

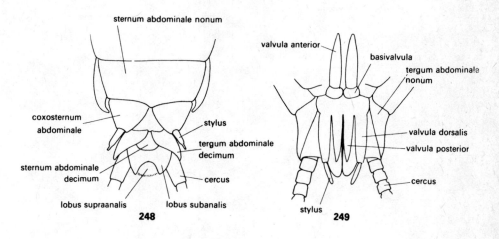

Fig. 247. The abdominal tympanal organ and its close environs in Orthoptera (Mecostethus-type). – Figs. 248–249. Abdominal end of the male nymph in ventral, and female nymph in dorsal view (after Matsuda 248–249)

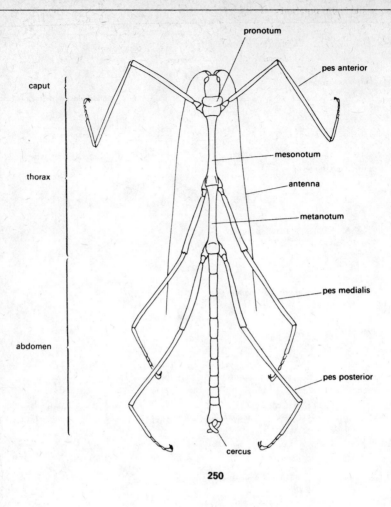

Fig. 250. Schematic drawing of the Phasmoidea in dorsal view

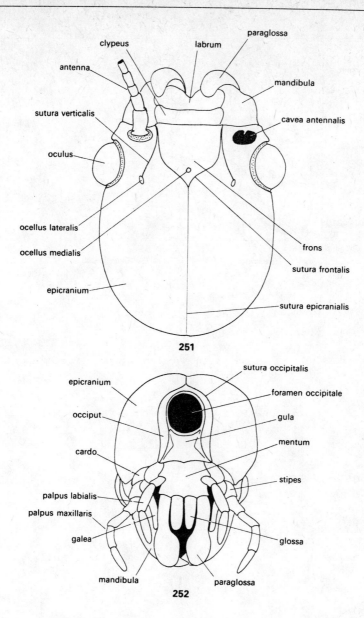

Figs. 251–252. Head of Phasmoidea in dorsal and ventral view

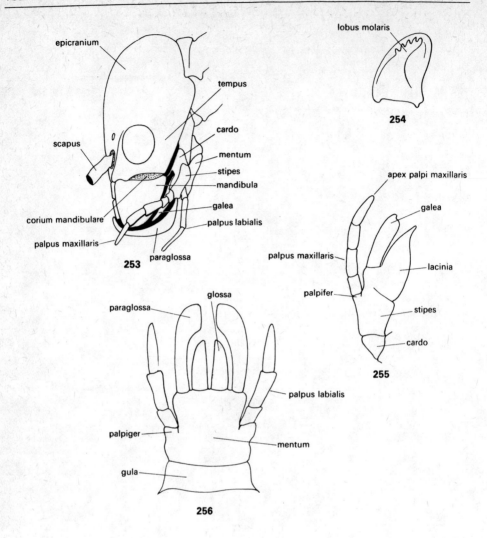

Fig. 253. Head of Phasmoidea in lateral view. – Fig. 254. The mandible. – Fig. 255. The maxilla. – Fig. 256. The labium

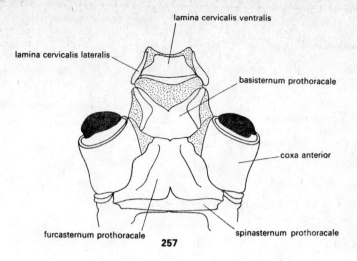

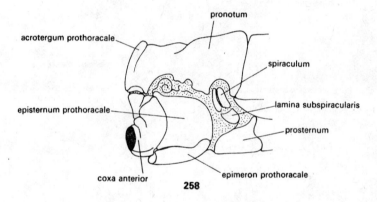

Figs. 257–258. Prothorax of Phasmoidea (Cyphocrania-type) in ventral and lateral view

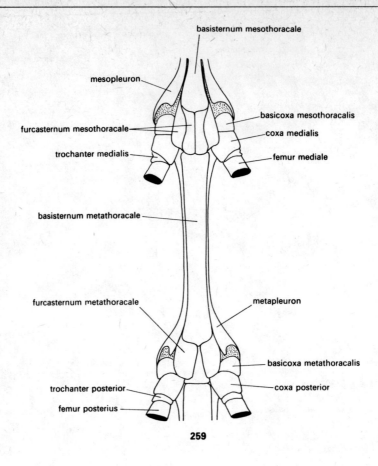

Fig. 259. Meso- and metathorax of Phasmoidea (Diapheromera-type) in ventral view

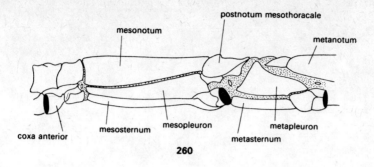

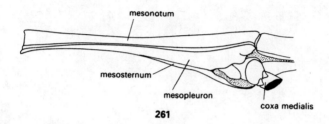

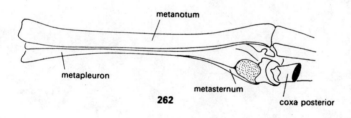

Fig. 260. Thorax of Phasmoidea (Cyphocrania-type) in lateral view. – Figs. 261–262. Meso- and metathorax of the Diapheromera-type in lateral view

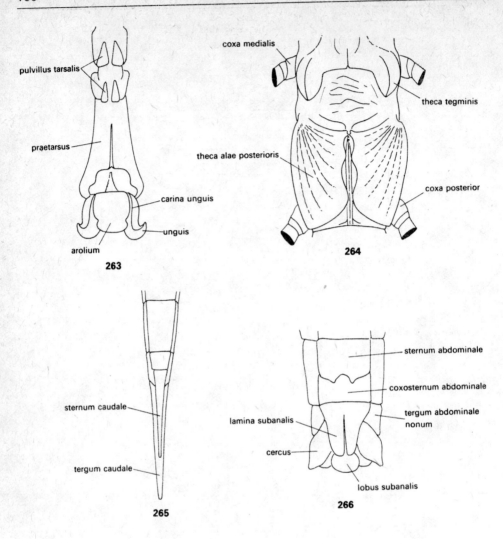

Fig. 263. Typical tarsus of Phasmoidea. – Fig. 264. The wing pads of Phasmoidea. – Figs. 265–266. Abdominal end in ventral view. – 265: Diapheromera, and 266: Cyphocrania

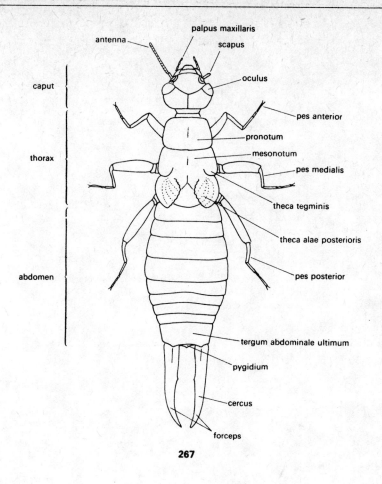

Fig. 267. Schematic drawing of the Dermaptera (Forficula-type) in dorsal view

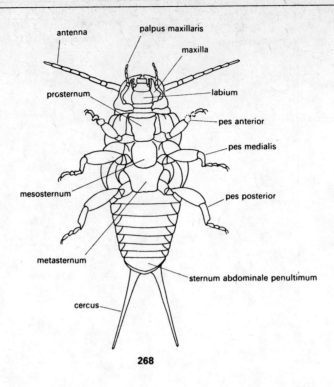

Fig. 268. Schematic drawing of the Dermaptera (Hemimerus-type) in ventral view

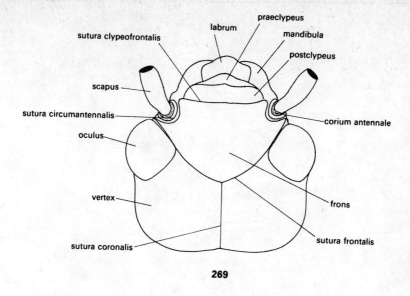

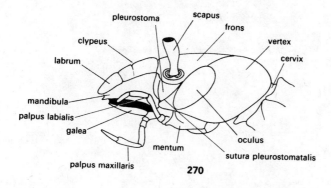

Figs. 269–270. Head of Dermaptera (Labidura- and Forficula-type) in dorsal and lateral view

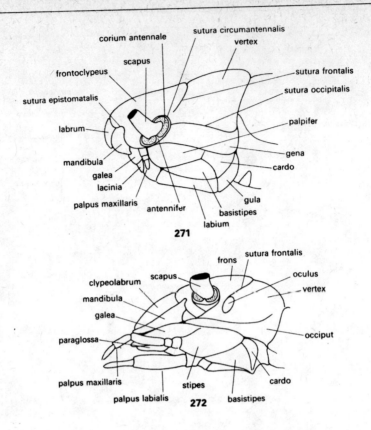

Figs. 271–272. Head of Dermaptera (Hemimerus- and Arixenia-type) in lateral view (after Giles)

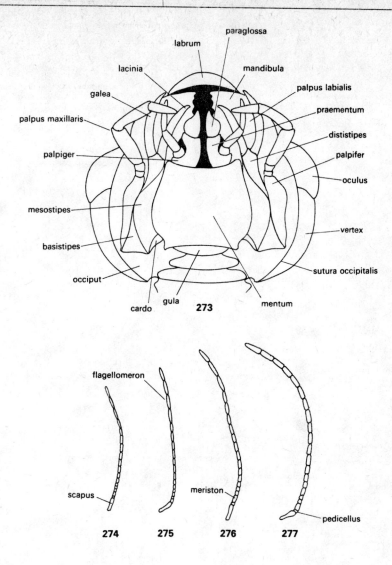

Fig. 273. Head of Dermaptera in ventral view. – Figs. 274–277. The antenna. – 274: third, 275: fourth, 276: fifth nymphal stage, and 277: adult (after Giles 274–277)

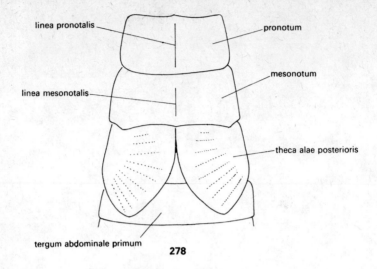
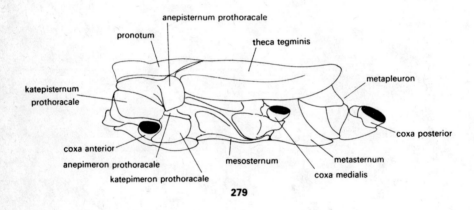

Figs. 278–279. Thorax of Dermaptera in dorsal and ventral view (after Günther and Herter 279)

Holomorphosis I: Paurometabolia

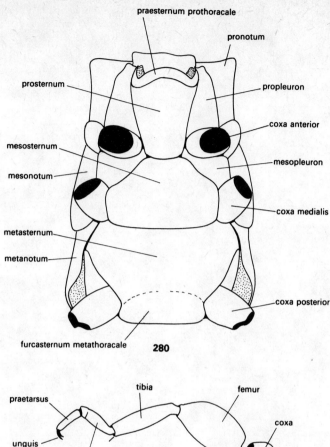

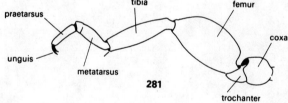

Fig. 280. Thorax of Dermaptera in ventral view. – Fig. 281. Typical thoracic leg

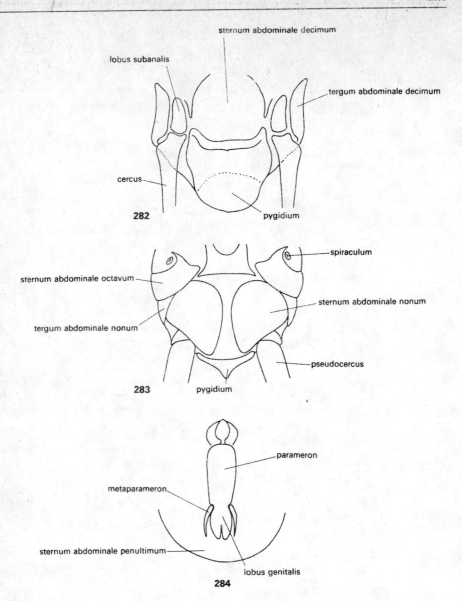

Figs. 282–283. Abdominal end of Dermaptera in ventral view. – 282: Hemimerus-type, and 283: Arixenia-type. – Fig. 284. The genital apparatus with the penultimate sternite of nymphal male from inside (after Matsuda 282–283)

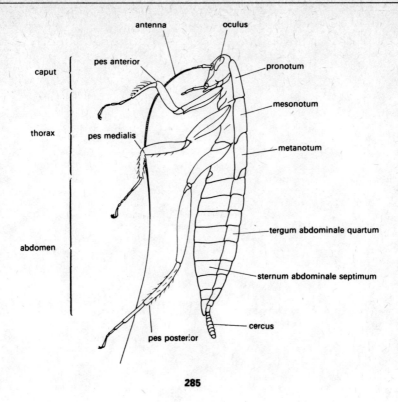

Fig. 285. Schematic drawing of the Blattaria in lateral view

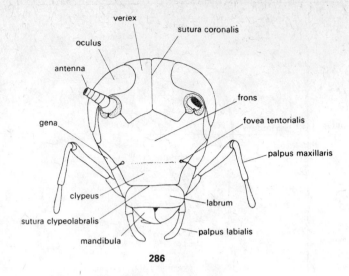

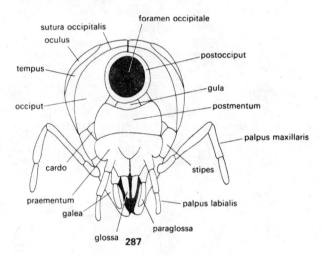

Figs. 286–287. Head of Blattaria in frontal and posterior view

Holomorphosis I: Paurometabolia

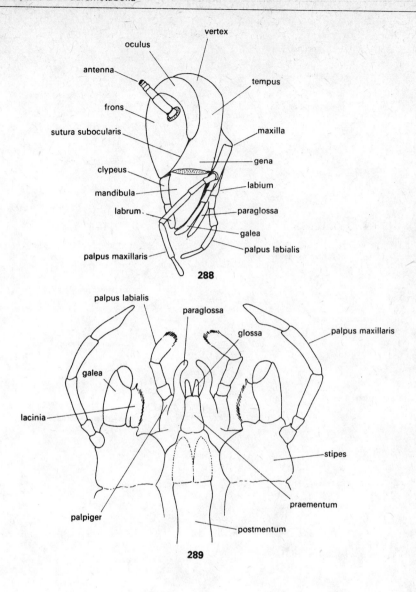

Fig. 288. Head of Blattaria in lateral view. – Fig. 289. The mouthparts in ventral view

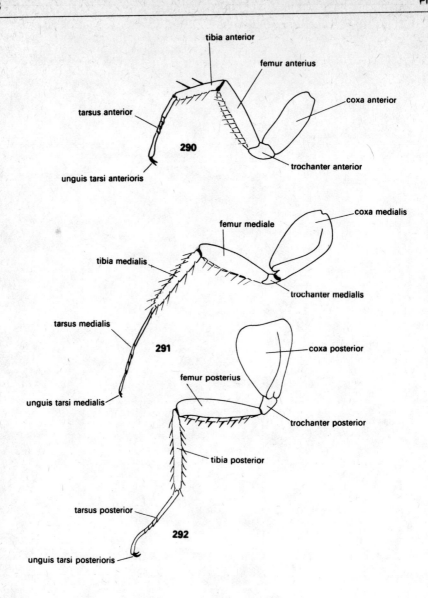

Figs. 290–292. Typical thoracic legs of Blattaria. –290: Fore, 291: middle, and 292: hind

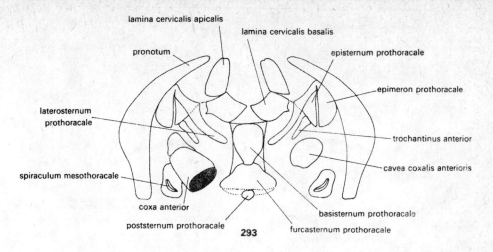
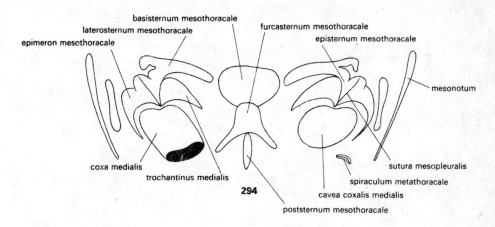

Figs. 293–294. Pro- and mesothorax of Blattaria in ventral view

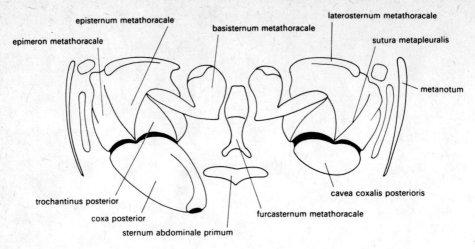

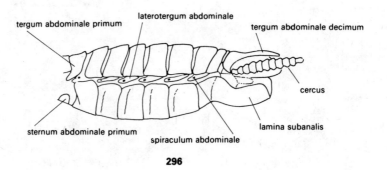

Fig. 295. Metathorax of Blattaria in ventral view. – Fig. 296. The abdomen in lateral view

Holomorphosis I: Paurometabolia

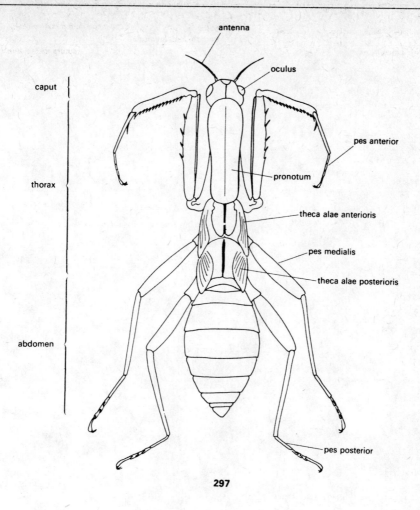

Fig. 297. Schematic drawing of the Mantoidea in dorsal view

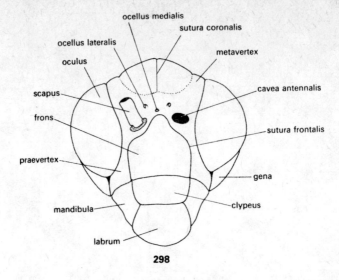

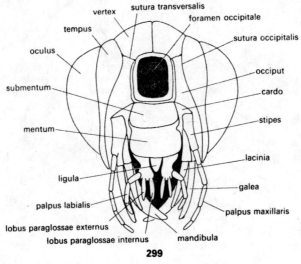

Figs. 298–299. Head of Mantoidea in frontal and posterior view

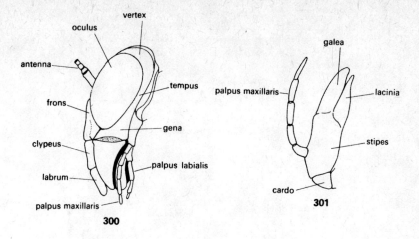
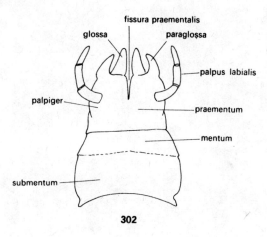

Fig. 300. Head of Mantoidea in lateral view. – Fig. 301. The maxilla. – Fig. 302. The labium

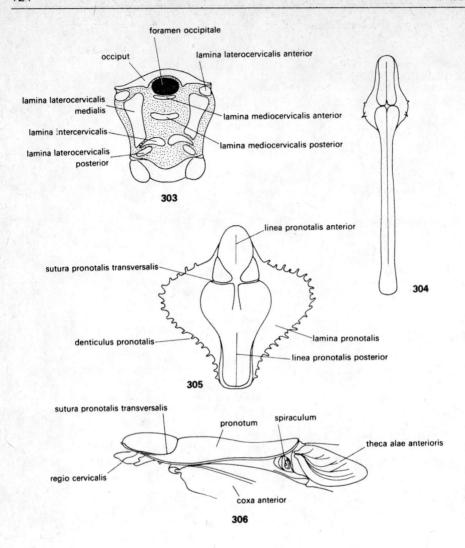

Fig. 303. The cervical region of Mantoidea in ventral view. – Figs. 304–306. Pronotum. – 304: Empusa, 305: Blepharopsis, and 306: Sphodromantis (after Beier 303)

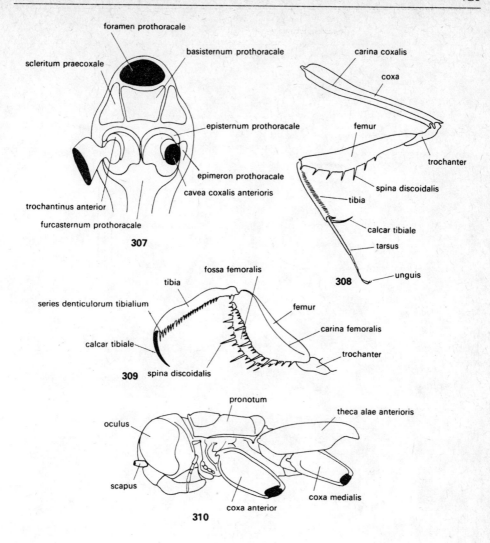

Fig. 307. Prosternum of Mantoidea in ventral view. – Figs. 308–309. The thoracic leg. – 308: Empusa-type, and 309: Blepharopsis-type. – Fig. 310. Head and thorax of the Eremiaphila-type

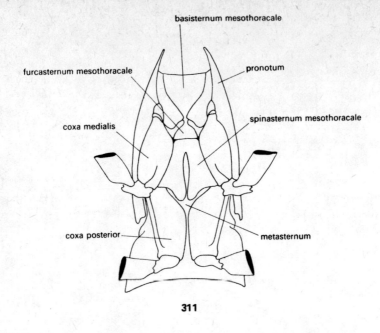

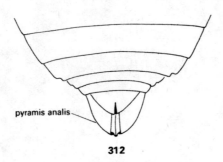

Fig. 311. Thorax of Mantoidea in ventral view. – Fig. 312. Abdominal end

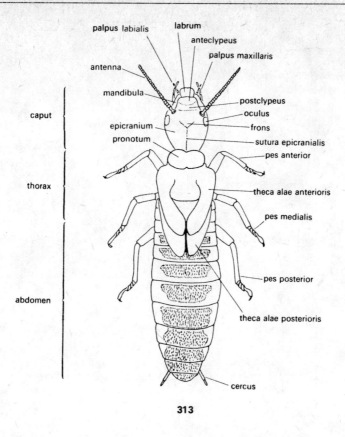

Fig. 313. Schematic drawing of the Isoptera in dorsal view

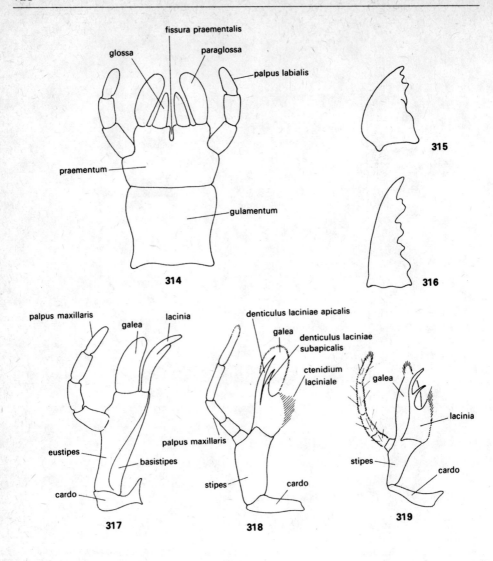

Figs. 314–319. Mouthparts of Isoptera. – 314: Labium, 315–316: mandibles, 317–319: maxillae, 317: typical form, 318: Termes-type, and 319: Eutermes-type (after Sjöstedt 318–319)

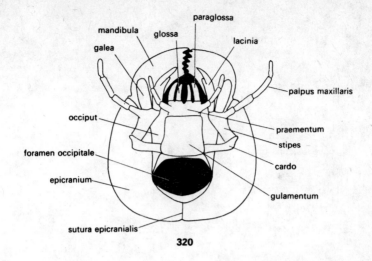

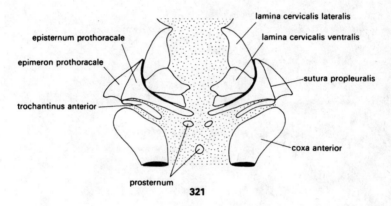

Fig. 320. Head of Isoptera in ventral view. – Fig. 321. The cervical region in ventral view (after Weidner 321)

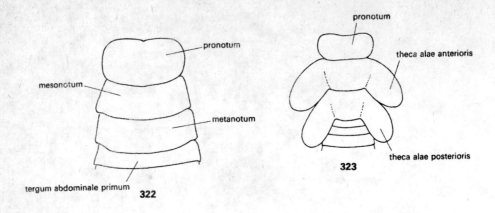

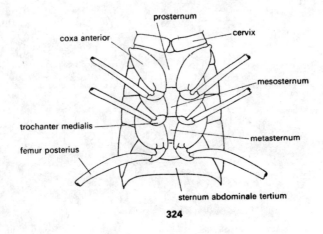

Figs. 322–323. Thorax of Isoptera without and with wing pads. – Fig. 324. The sternal region of the thorax in Hodotermes (after Weidner 324)

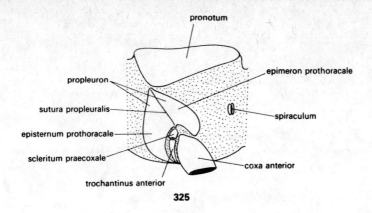

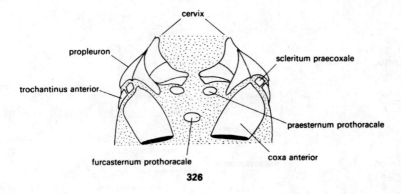

Figs. 325–326. Prothorax of Isoptera in lateral and ventral view (after Grassé, and Weidner, combined)

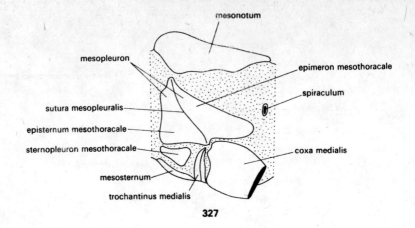
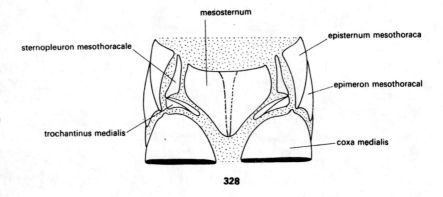

Figs. 327–328. Mesothorax of Isoptera in lateral and ventral view (after Grassé, and Weidner, combined)

Holomorphosis I: Paurometabolia

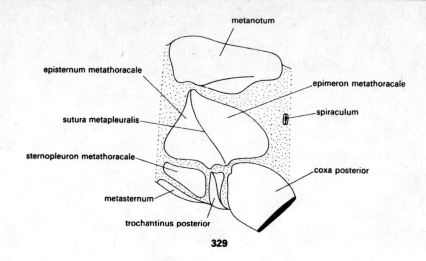

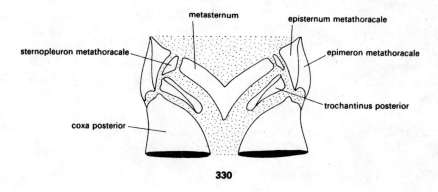

Figs. 329–330. Metathorax of Isoptera in lateral and ventral view (after Grassé, and Weidner, combined)

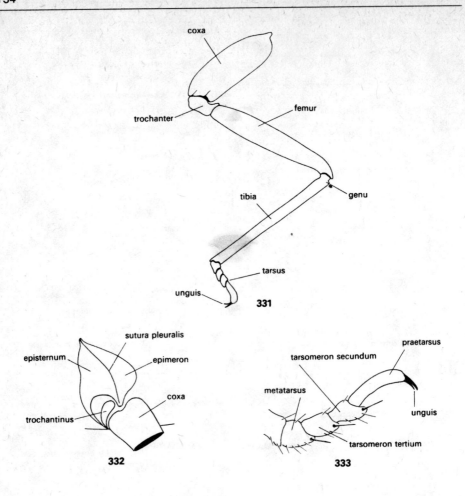

Fig. 331. Typical thoracic leg of Isoptera. – Fig. 332. The coxal region. – Fig. 333. Typical tarsus (after Weidner 332)

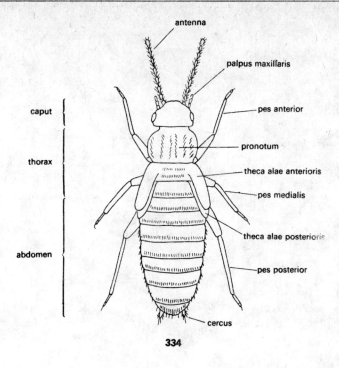

Fig. 334. Schematic drawing of the Zoraptera in dorsal view

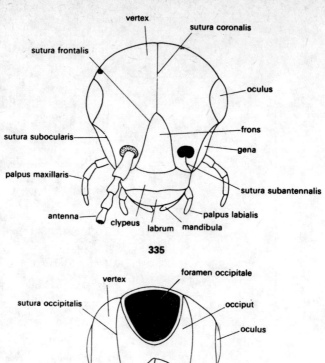

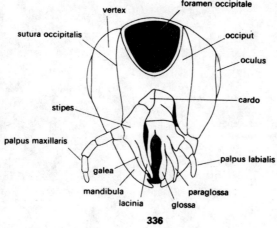

Figs. 335–336. Head of Zoraptera in frontal and posterior view

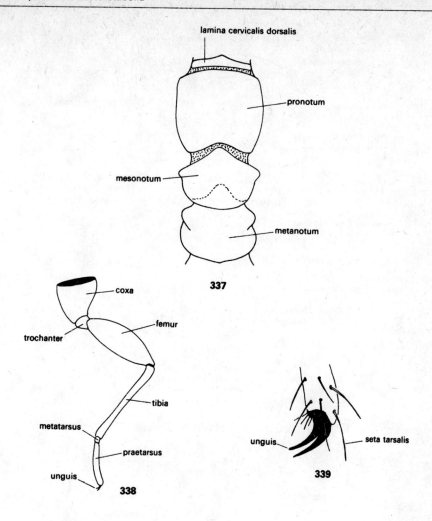

Fig. 337. Thorax of Zoraptera in dorsal view. – Fig. 338. Typical thoracic leg. – Fig. 339. The claw

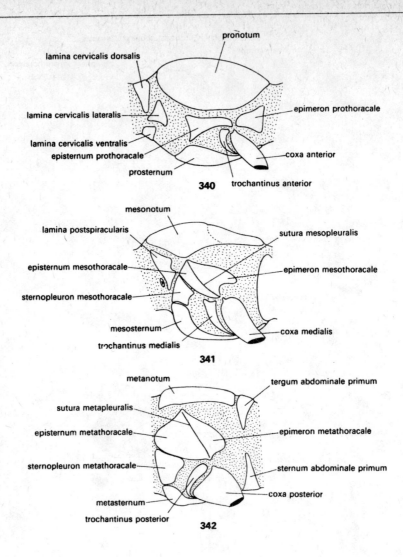

Figs. 340–342. Pro-, meso- and metathorax of Zoraptera in lateral view

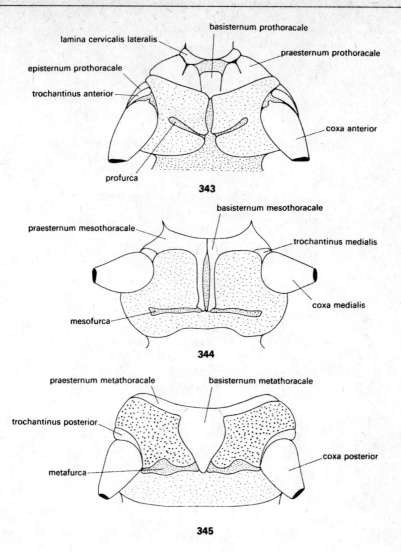

Figs. 343–345. Pro-, meso- and metathorax of Zoraptera in ventral view (after Matsuda, modified)

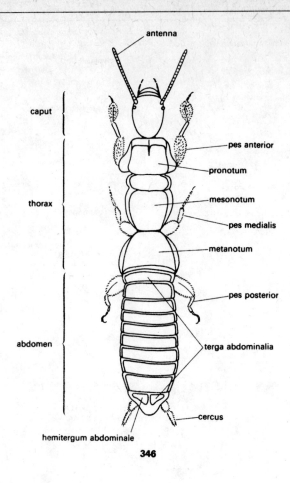

Fig. 346. Schematic drawing of the Embioptera in dorsal view (after Krauss)

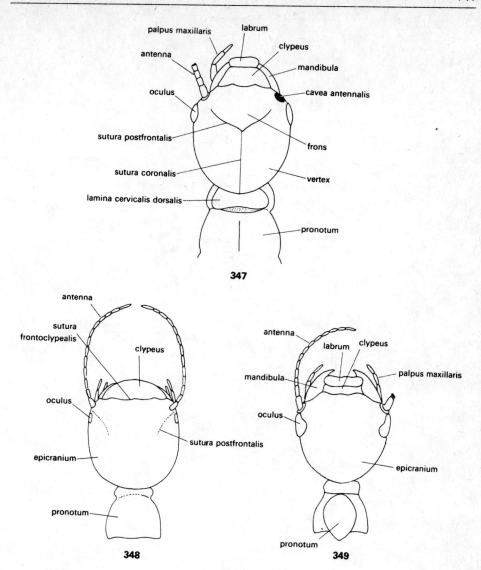

Figs. 347–349. Head and pronotum of Embioptera in dorsal view. – 347: Embia-type, 348: Gynembia-type, and 349: Teratembia-type (after Krauss 347, 349, and Peterson 348)

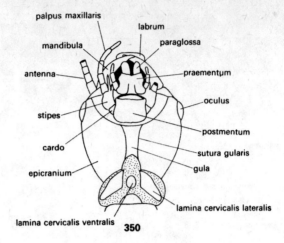
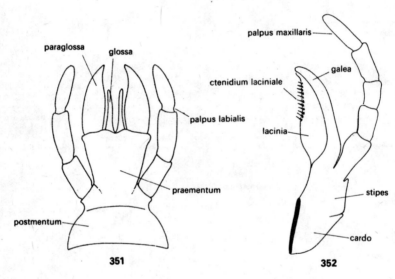

Fig. 350. Head of Embioptera in ventral view. – Fig. 351. The labium. – Fig. 352. The maxilla (after Rähle)

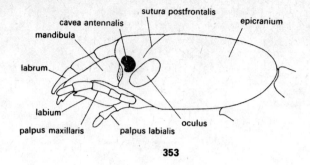

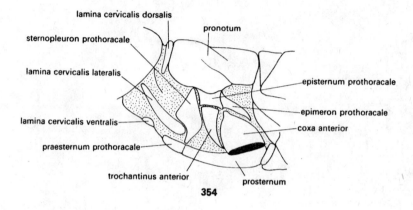

Fig. 353. Head of Embioptera in lateral view. – Fig. 354. Prothorax in lateral view (after Rähle)

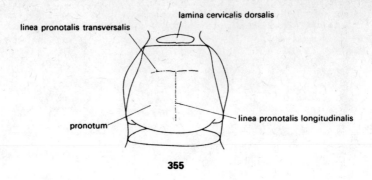

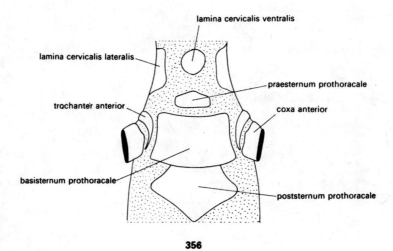

Figs. 355–356. Prothorax of Embioptera in dorsal and ventral view (after Krauss, and Matsuda, modified)

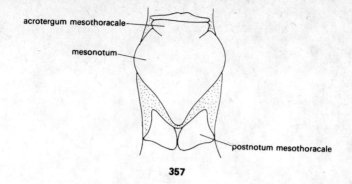

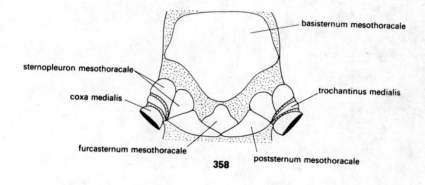

Figs. 357–358. Mesothorax of Embioptera in dorsal and ventral view (after Krauss, and Matsuda, modified)

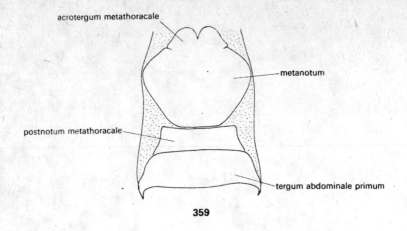

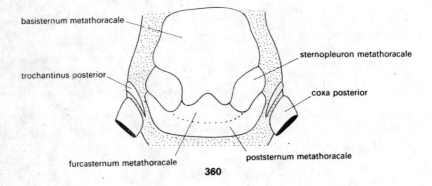

Figs. 359–360. Metathorax of Embioptera in dorsal and ventral view (after Krauss, and Matsuda, modified)

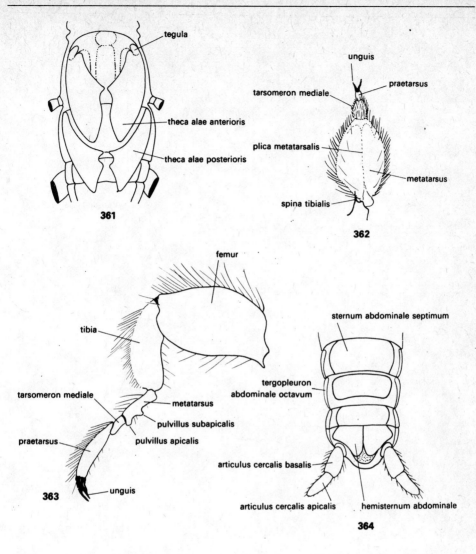

Fig. 361. Thorax of Embioptera with wing pads. – Fig. 362. The tarsus of the fore leg. – Fig. 363. Typical thoracic leg. – Fig. 364. Abdominal end of female nymph in ventral view (after Krauss 361–362, 364, and Westwood 363)

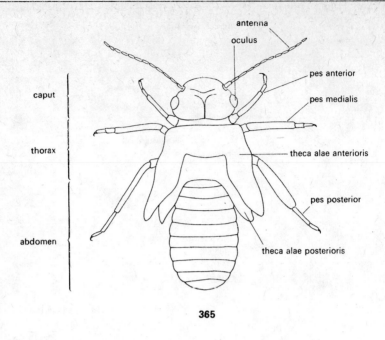

Fig. 365. Schematic drawing of the Psocoptera in dorsal view

Holomorphosis I: Parametabolia

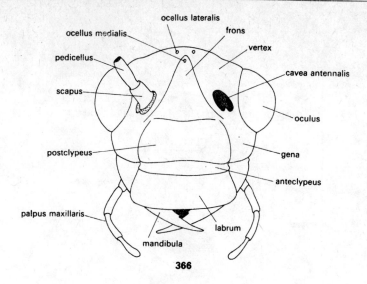

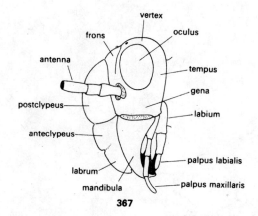

Figs. 366–367. Head of Psocoptera in frontal and lateral view

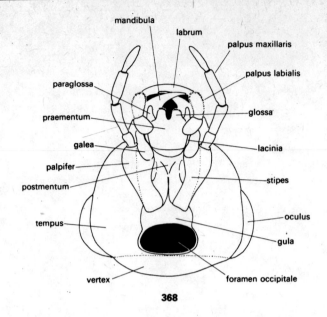
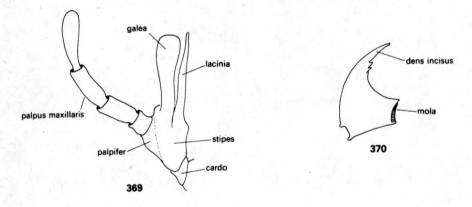

Fig. 368. Head of Psocoptera in ventral view. – Fig. 369. The maxilla. – Fig. 370. The mandible (after Weidner, modified 369–370)

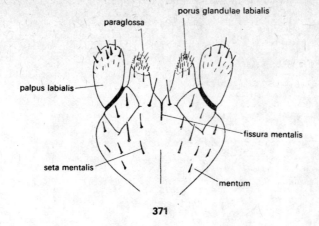

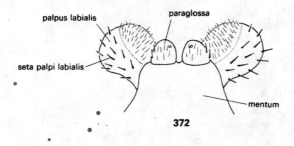

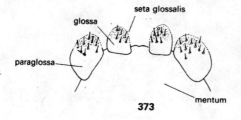

Figs. 371–373. Labium of Psocoptera. – 371: Lepinotus-type, 372: Psococerastis-type, and 373: Stenopsocus-type (after Weidner, modified)

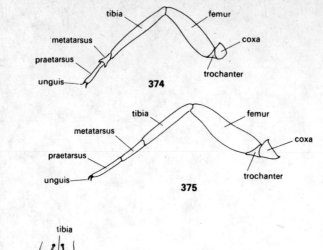
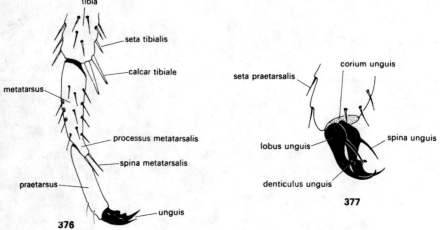

Figs. 374–375. Two types of thoracic leg of Psocoptera. — Fig. 376. Typical tarsus. – Fig. 377. The claw

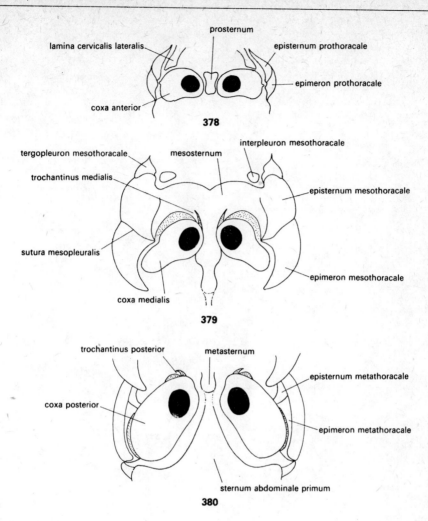

Figs. 378–380. Pro-, meso- and metathorax of Psocoptera in ventral view (after Matsuda, modified)

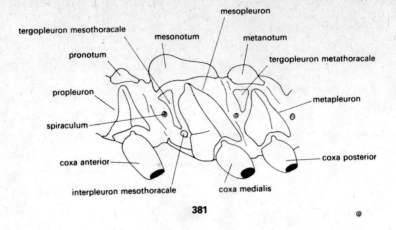

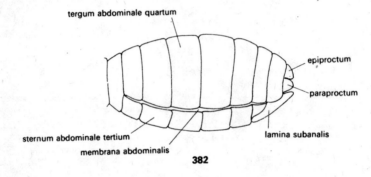

Fig. 381. Thorax of Psocoptera in lateral view. – Fig. 382. Abdomen in lateral view

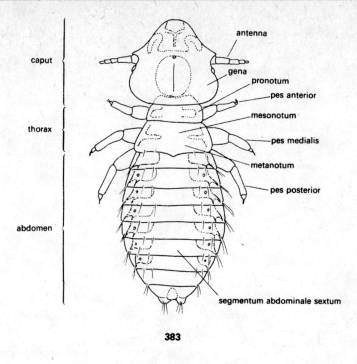

Fig. 383. Schematic drawing of the Mallophaga in dorsal view (after Peterson)

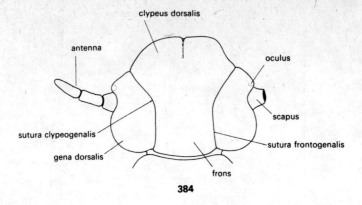

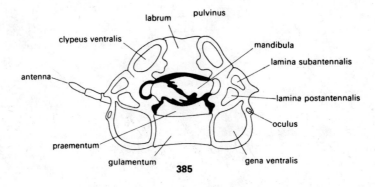

Figs. 384–385. Head of Mallophaga in dorsal and ventral view (after Buckup)

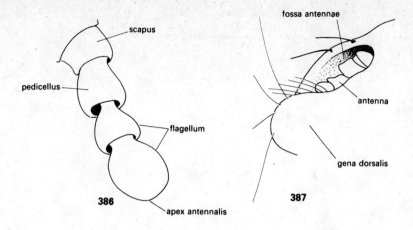

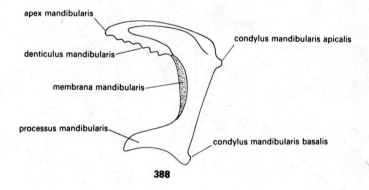

Fig. 386. Typical antenna of Mallophaga. – Fig. 387. The antennal furrow concealing the antenna. – Fig. 388. The mandible (after Buckup)

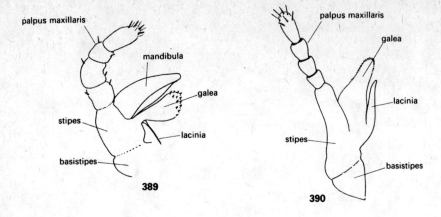

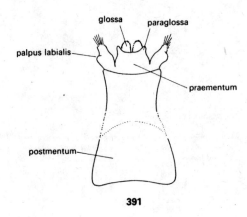

Figs. 389–390. The maxilla of Mallophaga. – 389: Myrsidea-type, and 390: Trimenopon-type. – Fig. 391. The labium (after Buckup)

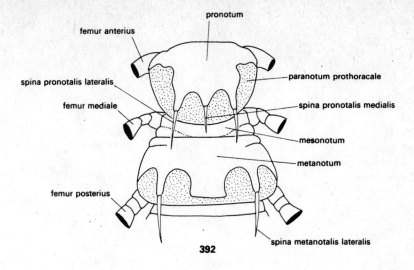

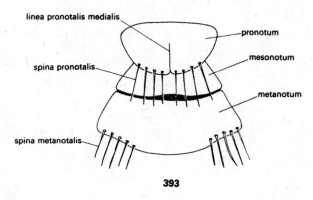

Figs. 392–393. Thorax of Mallophaga in dorsal view. – 392: Trimenopon-type, and 393: Myrsidea-type (after Mayer, modified)

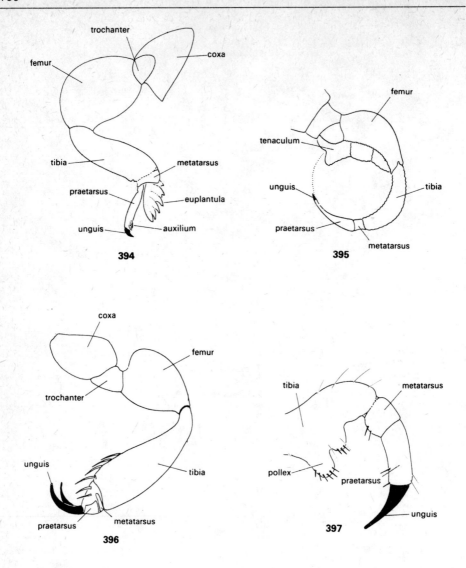

Figs. 394–397. Typical thoracic leg of Mallophaga. – 394: Myrsidea, 395: Gyropus, 396: Columbicola, and 397: Macrogyropus (after Kéler 394–395, Mayer 396, and Grassé 397)

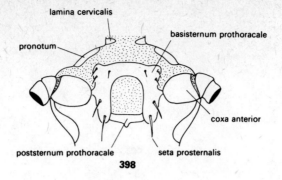
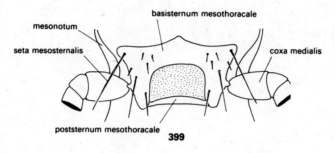
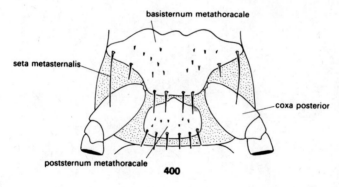

Figs. 398–400. Pro-, meso- and metathorax of Mallophaga (Trimenopon-type) in ventral view (after Mayer, modified)

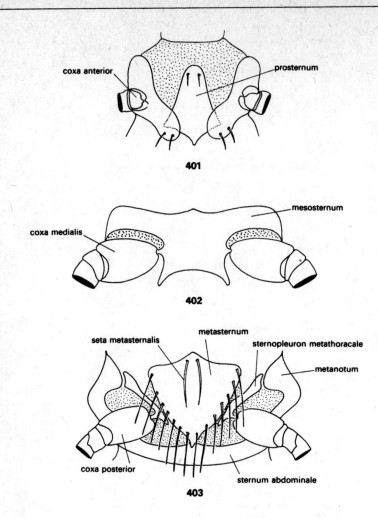

Figs. 401–403. Pro-, meso- and metathorax of Mallophaga (Myrsidea-type) in ventral view (after Matsuda, modified)

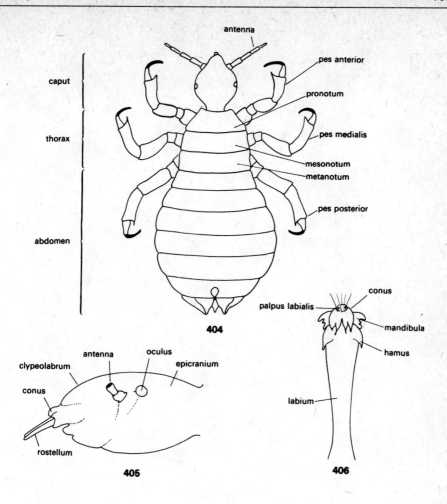

Fig. 404. Schematic drawing of the young nymph of Anoplura in dorsal view. – Fig. 405. Head in lateral view. – Fig. 406. The rostellum (after Grassé 405, and Wéber 406)

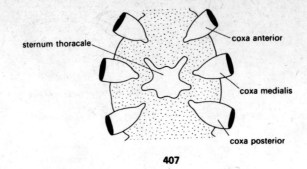

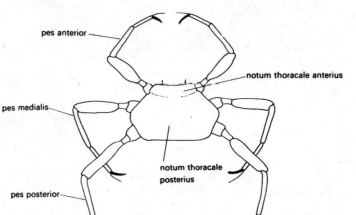

Fig. 407. The thoracic sternum of Anoplura. – Fig. 408. The thorax of the Haematomyzus-type with legs in dorsal view (after Matsuda 407, and Shvanvich 408)

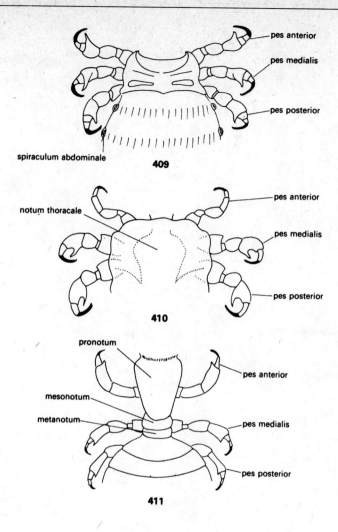

Figs. 409–411. Thorax of Anoplura with legs in dorsal view. – 409: Haematopinus-type, 410: Phthirus-type, and 411: Microthoracius-type (after Grassé, modified)

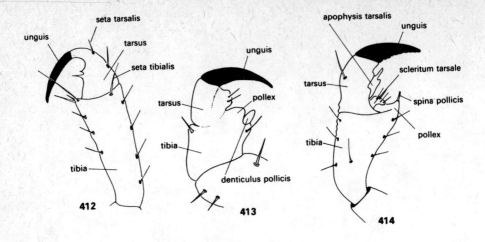

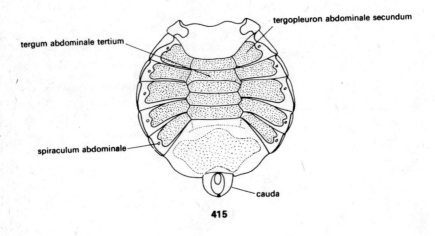

Figs. 412–414. Tibia and tarsus of Anoplura. – 412: Pediculus-type, 413: Antarctophthirius-type, and 414: Haematopinus-type. – Fig. 415. Abdomen in dorsal view (after Shvanvich)

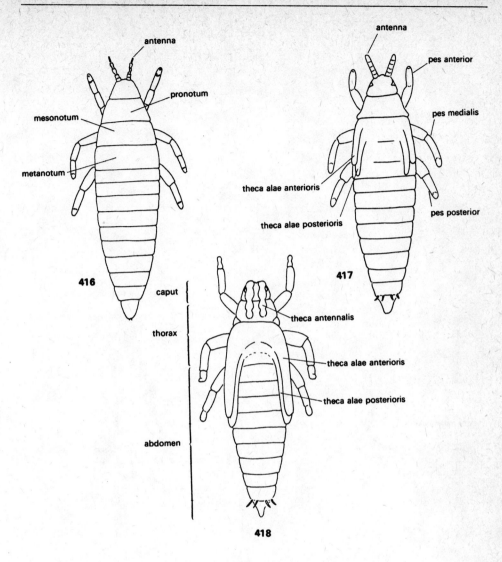

Figs. 416–418. Developmental stages of Thysanoptera in dorsal view. – 416: second nymph, 417: subpupa, and 418: pupa (after Priesner)

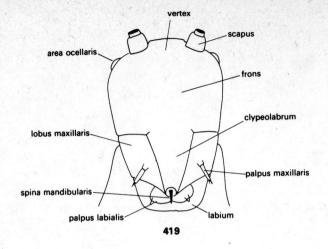
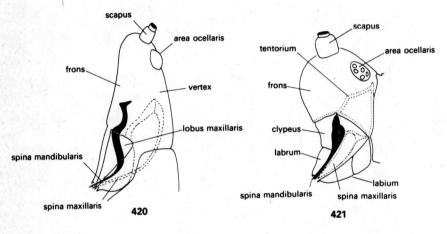

Figs. 419–421. Head of Thysanoptera in frontal and lateral view. – 419–420: Cephalothrips, and 421: Heliothrips (after Grassé)

Holomorphosis I: Parametabolia

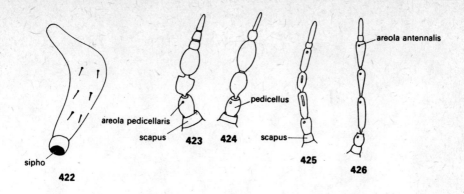

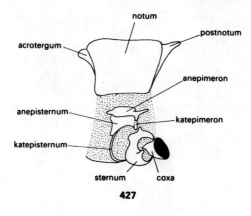

Fig. 422. The clypeolabrum of Thysanoptera in frontal view. – Fig. 423–426. The antenna of the first and second nymphal stage of Thrips and Heliothrips. – Fig. 427. Typical thoracic segment in lateral view (after Grassé 422, Priesner 423–426, and Mickoleit 427)

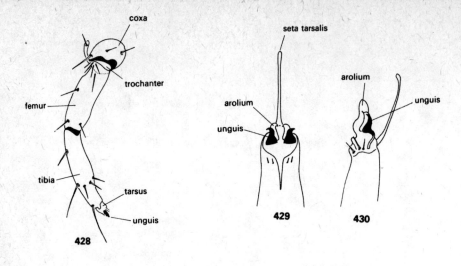
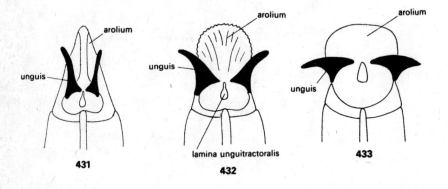

Fig. 428. Typical thoracic leg of Thysanoptera. – Figs. 429–433. Various types of tarsal end. – 429–430: Liothrips in ventral and lateral view, and 431–433: the arolium of Trichothrips in repose, in half and fully expanded state (after Grassé)

Holomorphosis I: Parametabolia

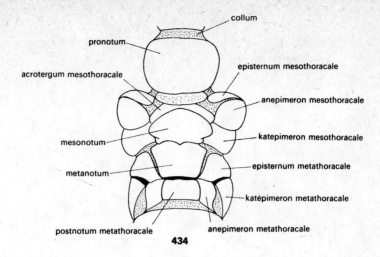

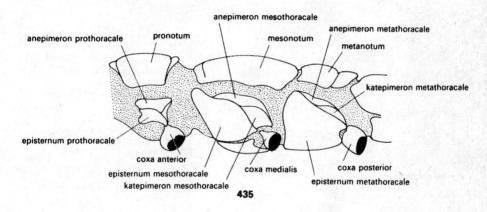

Figs. 434–435. Thorax of Thysanoptera in dorsal and lateral view

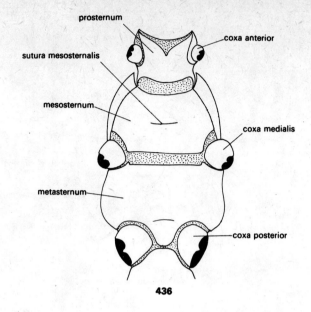
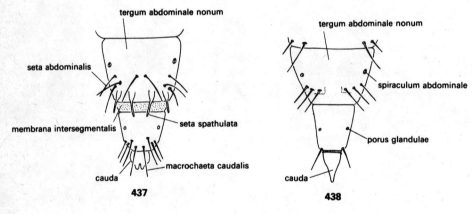

Fig. 436. Thorax of Thysanoptera in ventral view. – Figs. 437–438. Abdominal end in dorsal view. – 437: Aelothrips nymph, and 438: Neobeegeria pronymph (after Priesner 437–438)

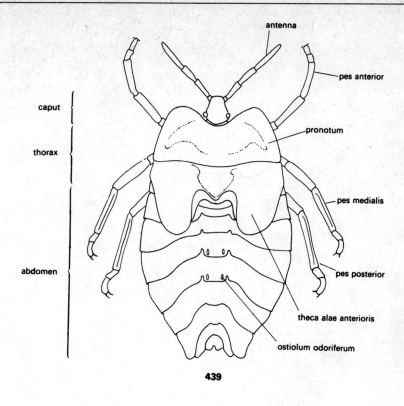

Fig. 439. Schematic drawing of the Heteroptera in dorsal view

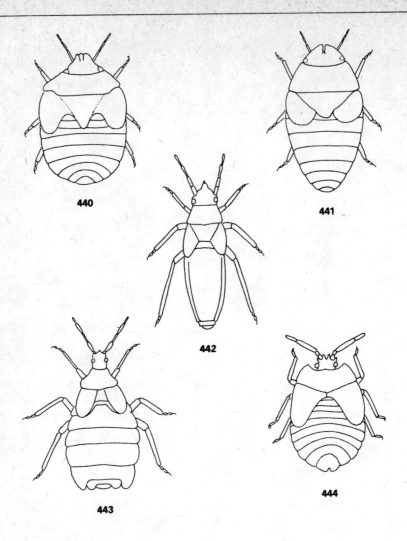

Figs. 440–444. Various types of terrestrial Heteroptera. – 440: Eurygaster, 441: Aelia, 442: Myrmus, 443: Enoplops, and 444: Acalypta (after Butler)

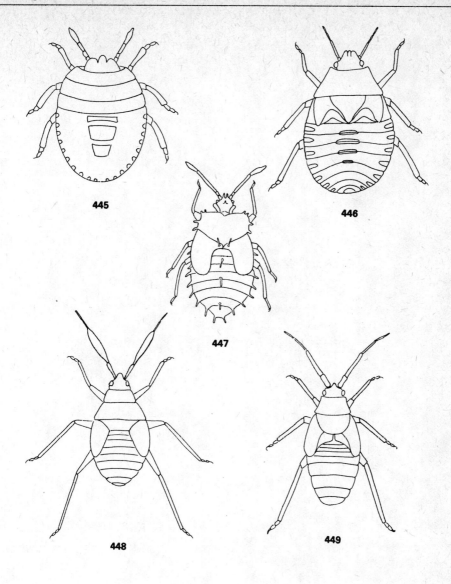

Figs. 445–449. Various types of terrestrial Heteroptera. – 445: Palomena, 446: Picromerus, 447: Monanthia, 448: Heterotoma, and 449: Pantilius (after Butler 445–446, 448–449, and Štusák 447)

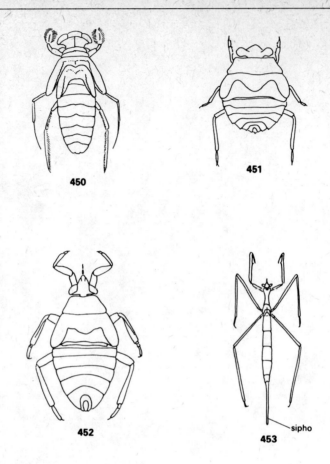

Figs. 450–453. Various types of aquatic Heteroptera. – 450: Corixa, 451: Gelastocoris, 452: Belostoma, and 453: Ranatra (after Peterson)

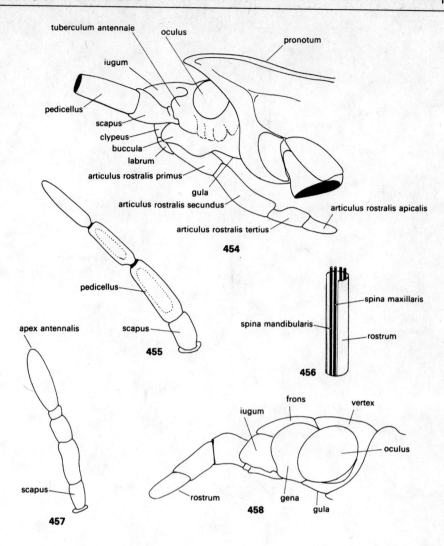

Fig. 454. Head of Heteroptera (Eusthenes-type) in lateral view. – Fig. 455: Antenna of Eusthenes. – Fig. 456. Part of rostrum with two pairs of stylets. – Fig. 457. Antenna of Alienates. – Fig. 458. Head of Heteroptera (Ranatra-type) in lateral view (after Vásárhelyi 457)

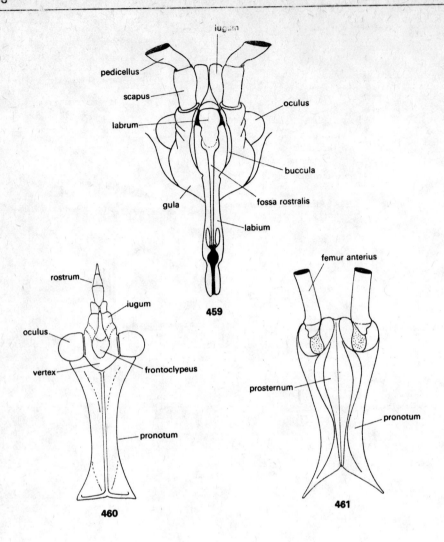

Fig. 459. Head of Heteroptera (Eusthenes-type) in ventral view. – Fig. 460. Head and pronotum of Ranatra. – Fig. 461. Prothorax of Ranatra in ventral view

Holomorphosis I: Parametabolia

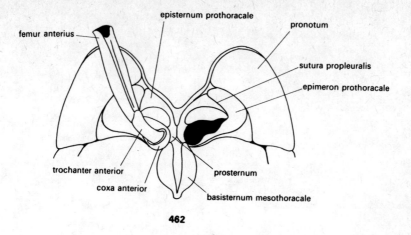

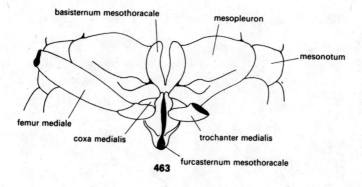

Figs. 462–463. Pro- and mesothorax of Heteroptera (Eusthenes-type) in ventral view

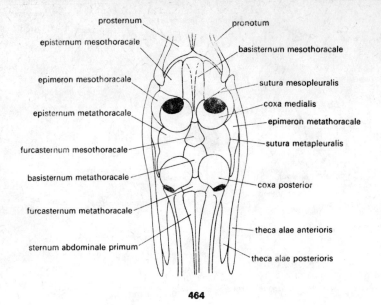

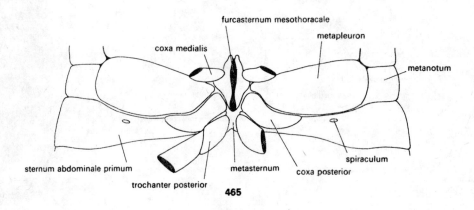

Fig. 464. Thorax of Heteroptera (Ranatra-type) in ventral view. – Fig. 465. Metathorax of the Eusthenes-type in ventral view

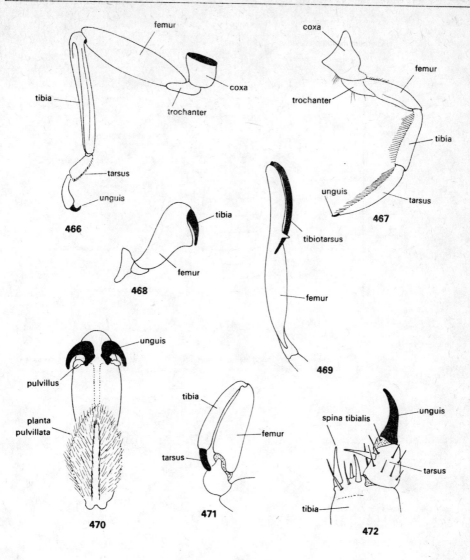

Figs. 466–472. Thoracic leg of Heteroptera. – 466: Terrestrial type, 467: aquatic type, 468: Macrocephalus-type, 469: Ranatra-type, 470: Eusthenes-type, 471: Nepa-type, and 472: Alienates-type (after Beier 466–467, Vásárhelyi 472)

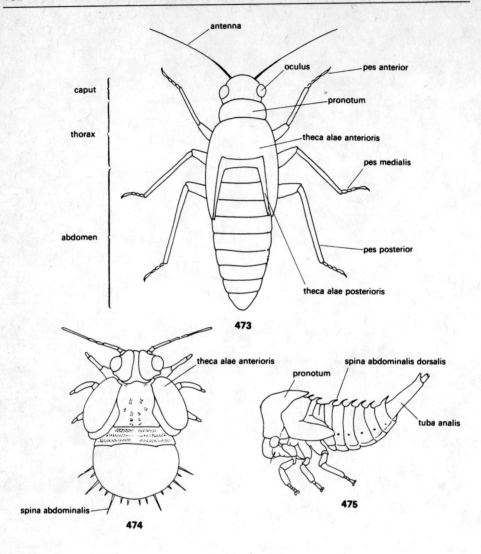

Fig. 473. Schematic drawing of the Homoptera in dorsal view. – Figs. 474–475. Two types of Homoptera. – 474: Psylla, and 475: Membracid type (after Grassé 474, and Peterson 475)

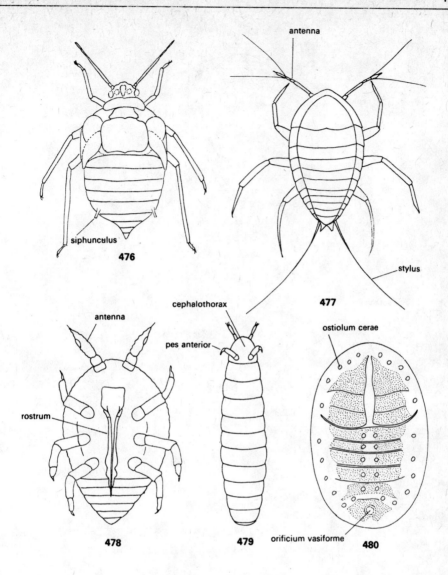

Figs. 476–480. Various developmental stages of Homoptera. – 476: Aphis nymph in dorsal, 477: Laccifer nymph in dorsal, 478: Phylloxera nymph in ventral, 479: Margarodes young nymph in ventral and 480: Aleuroparadoxus pupa in ventral view (after Imms 476, Grassé 477–480)

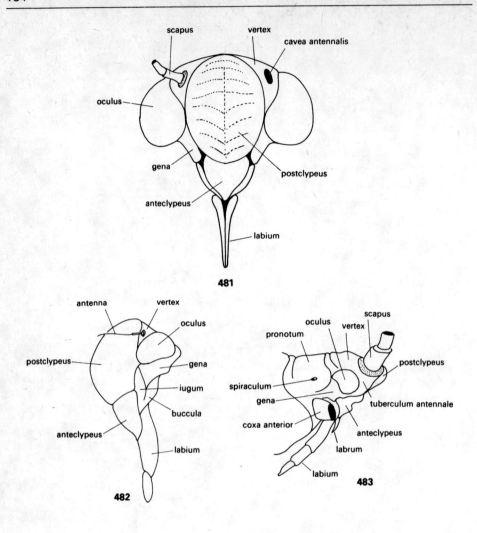

Figs. 481–483. Head of Homoptera. – 481: Cicada in frontal, 482: Tomaspis, and 483: Macrosiphon in lateral view (after Grassé 482–483)

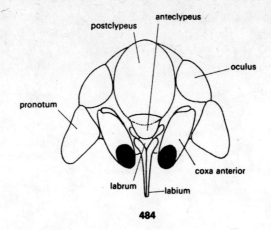
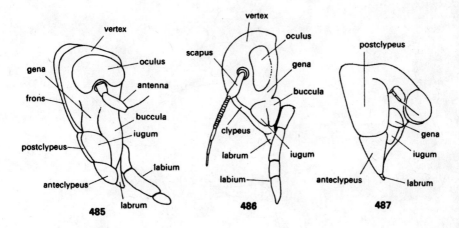

Figs. 484–487. Head of Homoptera. – 484: Cicada (with pronotum) in anterior, 485: Stenocranus, 486: Aleyrodes, and 487: Ulopa in lateral view (after Grassé 485–487)

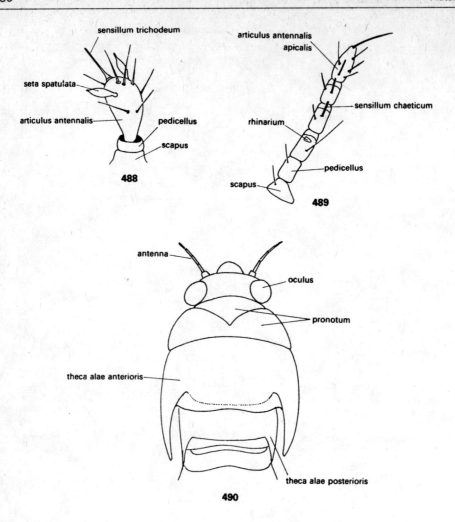

Figs. 488–489. Two types of antenna of Homoptera. – 488: Margarodes, and 489: Coccoid type. – Fig. 490. Head and thorax of Cicada in dorsal view (after Grassé 488, and Williams and Kosztarab 489)

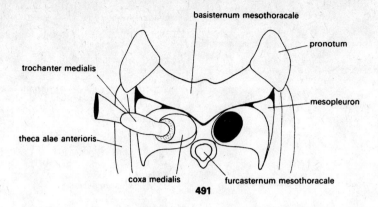
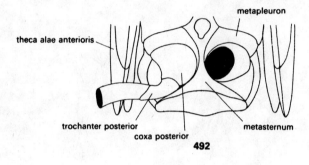

Figs. 491–492. Meso- and metathorax of Homoptera in ventral view

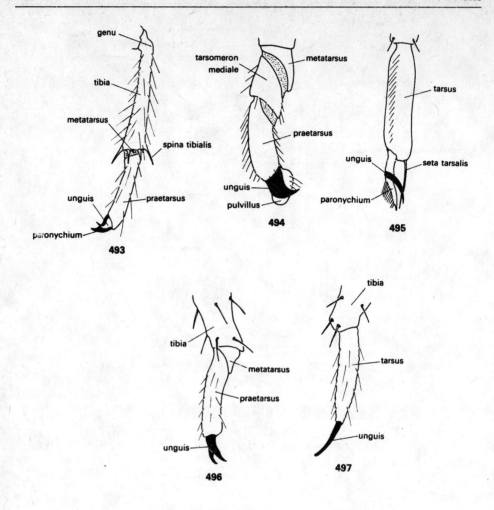

Figs. 493–497. Typical legs of Homoptera. – 493: Cicada, 494: Triecophora, 495: Aleurodes, 496: Aphis, and 497: Pseudococcus (after Grassé 494–497)

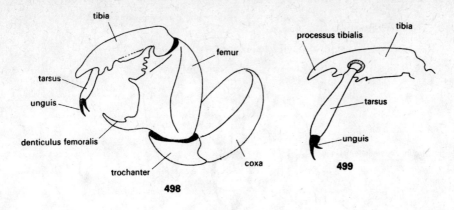

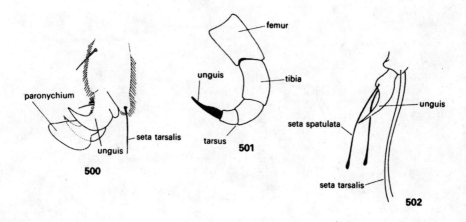

Figs. 498–502. Leg and parts of leg of Homoptera. – 498–499: Fore leg and tibia with tarsus of Cicada, 500: pretarsus of Aleurochiton, 501: leg of Margarodes, and 502: tarsal end of Chaetococcus (after Grassé 500–501, and Yang and Kosztarab 502)

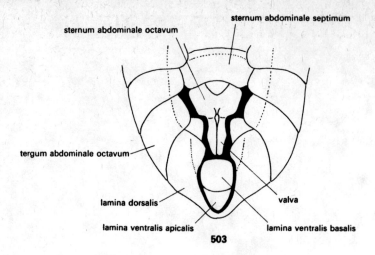

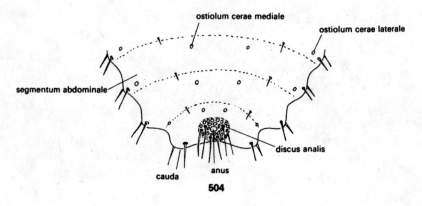

Figs. 503–504. Abdominal end of Homoptera in ventral view. – 503: Cicada-type, and 504: Antonina-type (after Yang and Kosztarab 504)

HOLOMORPHOSIS II.

Holomorphosis II: General part

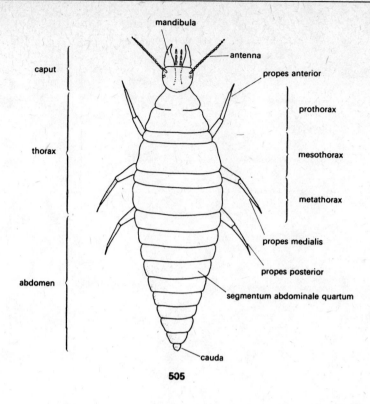

Figs. 505. Schematic drawing of the campodeiform larva in dorsal view

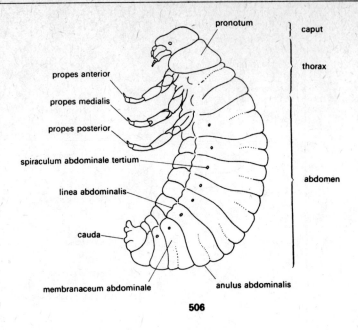

Fig. 506. Schematic drawing of the oligopod larva in lateral view

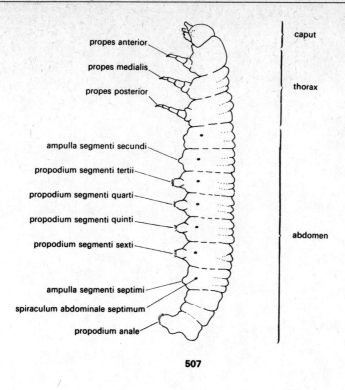

Fig. 507. Schematic drawing of the polypod larva in lateral view

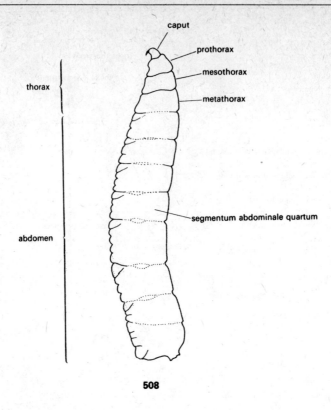

Fig. 508. Schematic drawing of the apod larva in lateral view

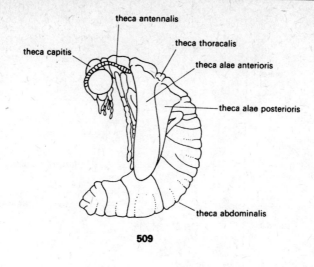
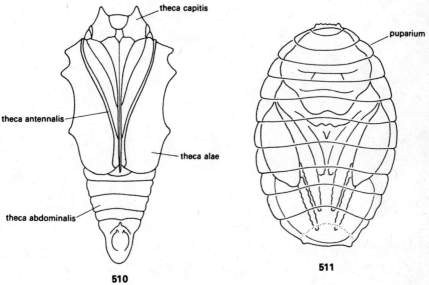

Figs. 509–511. Three types of pupa. – 509: Free (pupa libera), 510: obtected (pupa obtecta), and 511: coarctate (pupa coarctata).

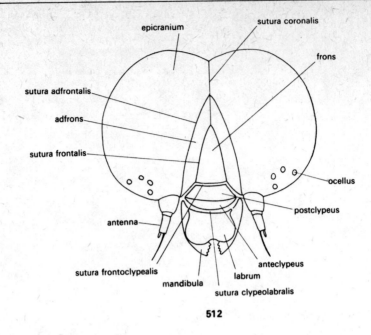

Fig. 512. Schematic drawing of the larva head in frontal view

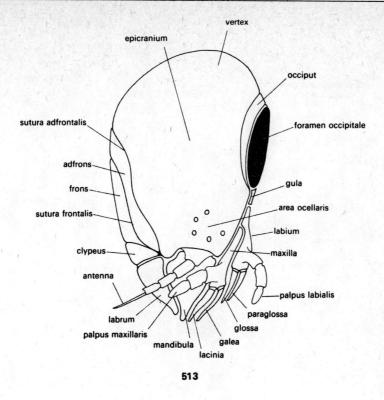

Fig. 513. Schematic drawing of the larva head in lateral view

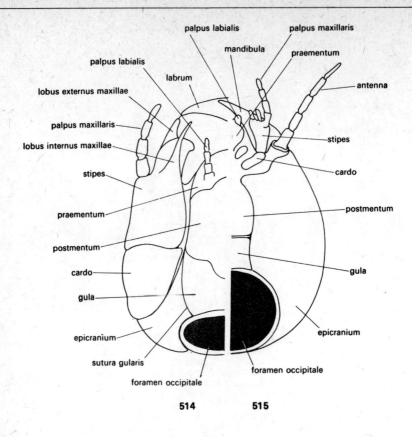

Figs. 514–515. Schematic drawing of two forms of the larva head in ventral view

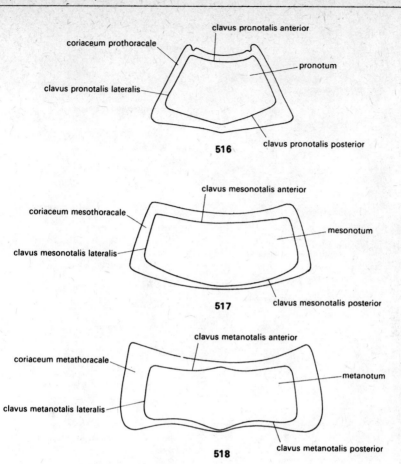

Figs. 516–518. Pro-, meso- and metathorax schematically in dorsal view

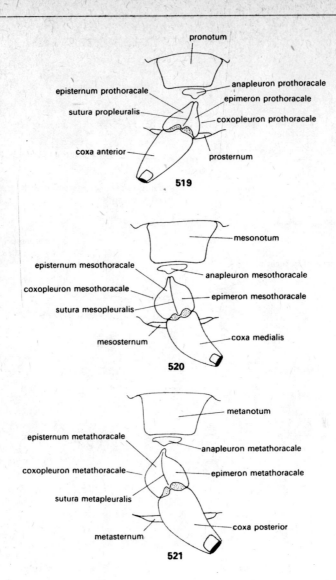

Figs. 519–521. Pro-, meso- and metathorax schematically in lateral view

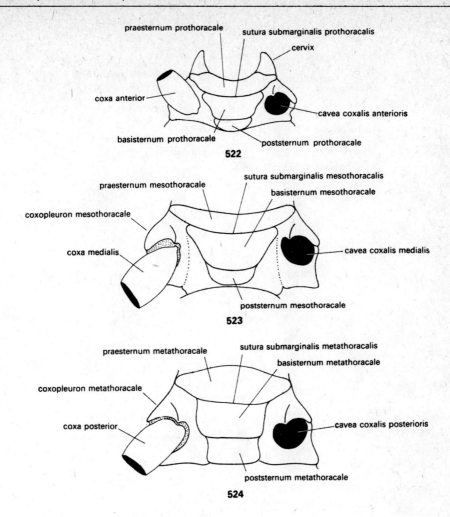

Figs. 522–524. Pro-, meso- and metathorax schematically in ventral view

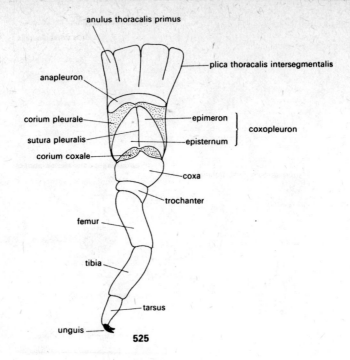
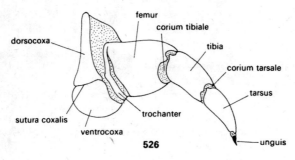

Fig. 525. Typical thoracic segment (segmentum thoracale) with proleg (propes) schematically in lateral view. – Fig. 526. Thoracic leg

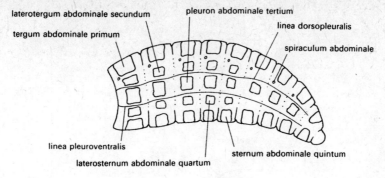

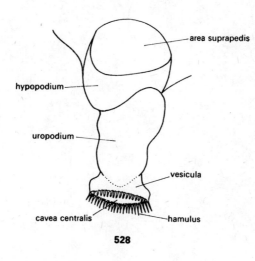

Fig. 527. Schematic drawing of the hypothetical sclerotization of the abdomen in lateral view. –
Fig. 528. Abdominal leg (propodium)

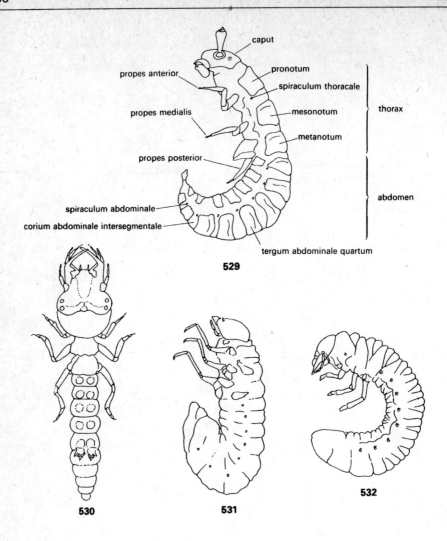

Fig. 529. Schematic drawing of the Coleoptera in lateral view. – Figs. 530–532. Characteristic types of Coleoptera larva. – 530: Cicindela in dorsal, 531: Chlamys, and 532: Phyllophaga in lateral view (after Peterson 530–532)

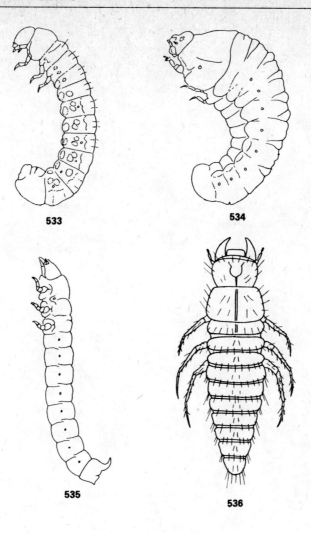

Figs. 533–536. Characteristic types of Coleoptera larva. – 533: Byrrhus in lateral, 534: Xylobiops in lateral, 535: Phellopsis in lateral, and 536: Phodaga in dorsal view (after Peterson 533–535, and MacSwain 536)

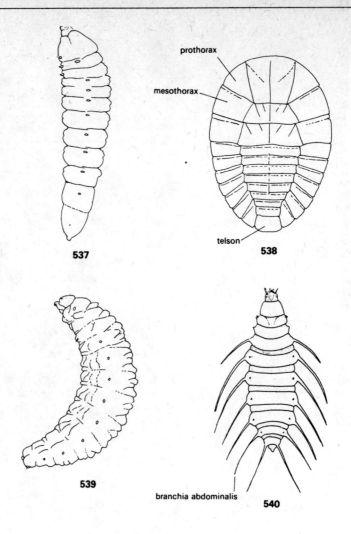

Figs. 537–540. Characteristic types of Coleoptera larva. – 537: Orthosoma in lateral, 538: Psephenus in dorsal, 539: Curculio in lateral, and 540: Berosus in dorsal view (after Peterson)

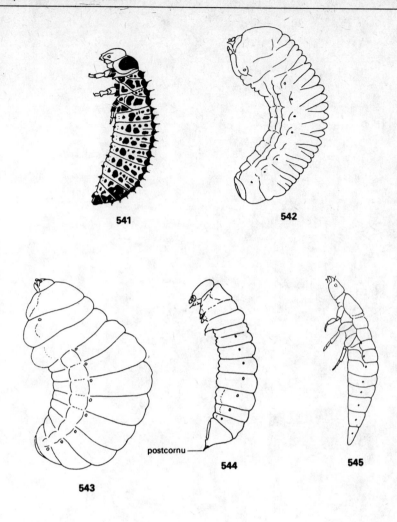

Figs. 541–545. Characteristic types of Coleoptera larva. – 541: Coccinella, 542: Scolytus, 543: Zabrotes, 544: Tomoxia, and 545: Hydrocanthus in lateral view (after Ghilarov 541, and Peterson 542–545)

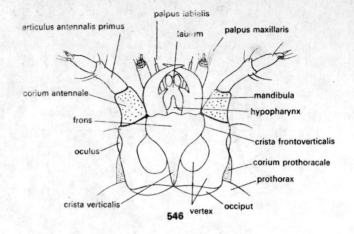

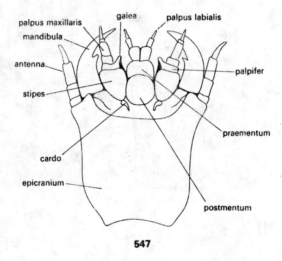

Figs. 546–547. Head of Coleoptera in dorsal and ventral view (after Meixner 546, and Peterson 547)

Holomorphosis II: Holometabolia

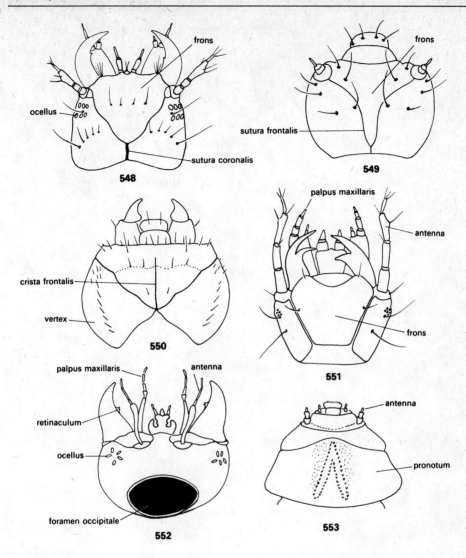

Figs. 548–553. Typical head of Coleoptera in dorsal view. – 548: Cymindis, 549: Oedionychis, 550: Anthophylax, 551: Badister, 552: Hydrous, and 553: Chalcophora (after Ghilarov 548, 551, and Peterson 549–550, 552–553)

14*

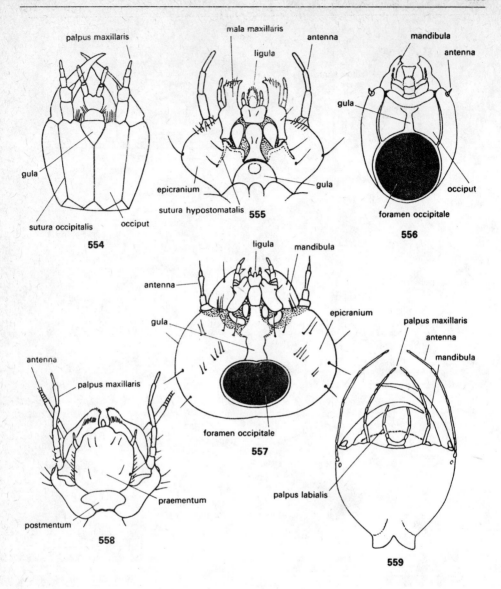

Figs. 554–559. Typical head of Coleoptera in ventral view. – 554: Hister, 555: Neopyrochroa, 556: Monochamus, 557: Tenebrio, 558: Cyphon, and 559: Dytiscus (after Ghilarov 554, and Peterson 555–559)

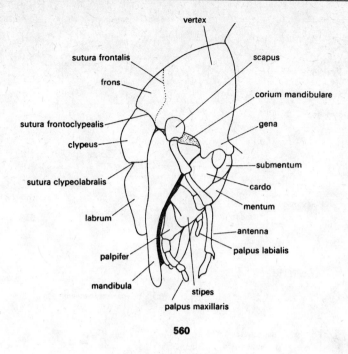

Fig. 560. Head of Coleoptera in lateral view

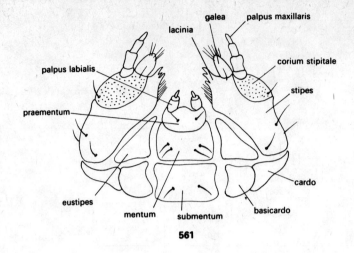
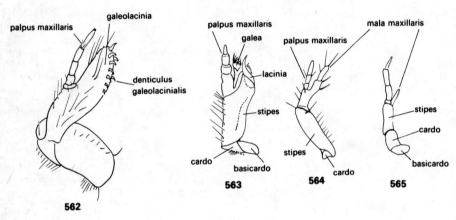

Figs. 561–565. Mouthparts of Coleoptera. – 561: The maxilla and labium of the Byrrhus-type. – 562–565: Four types of maxilla. – 562: Phyllophaga, 563: Dermestes, 564: Cicindela, and 565: Philonthus (after Ghilarov 561, and Peterson 562–565)

Holomorphosis II: Holometabolia

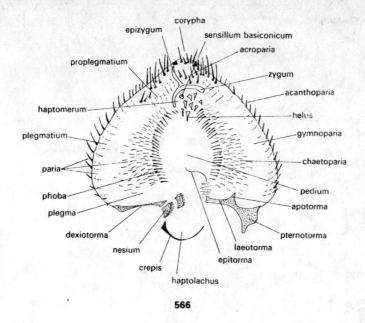

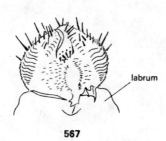

Figs. 566–567. Two forms of epipharynx of Coleoptera (after Peterson)

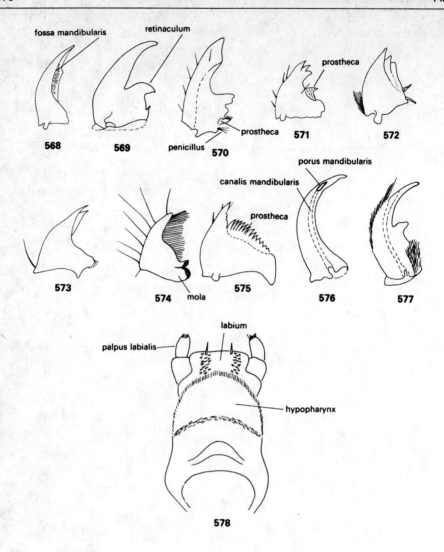

Figs. 568–577. Some characteristic types of the mandible of Coleoptera. – 568: Cantharis, 569: Calosoma, 570: Phyllophaga, 571: Oryzaephilus, 572: Dermestes, 573: Hippodamia, 574: Cyphon, 575: Cryptarcha, 576: Dineutes, and 577: Photuris. – Fig. 578. The labium and hypopharynx (after Peterson)

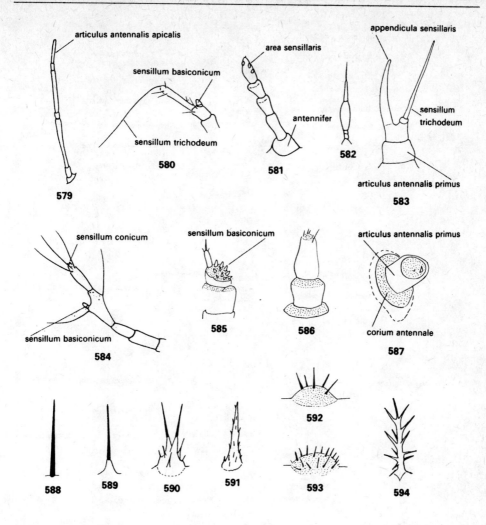

Figs. 579–587. Some characteristic types of the antenna of Coleoptera. – 579: Dytiscus, 580: Meloid type, 581: Scarabaeoid type, 582: Peltodytes, 583: Xestocis, 584: Staphylinid type, 585: Parallelostethus, 586: Epilachna, and 587: Physonota. – Figs. 588–594. Tegumentary formations. – 588: Seta, 589: chalaza, 590: parascolus, 591: sentus, 592: struma, 593: verruca, and 594: scolus (after Meixner 579–581, 584, and Peterson 582–583, 585–594)

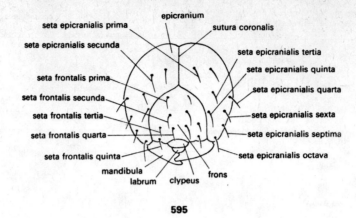

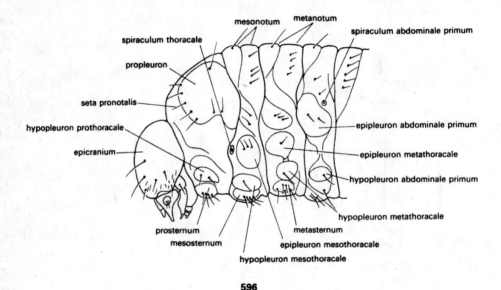

Figs. 595–596. The chaetotaxy of the head and the sclerotization of the thorax of Coleoptera (after Ghilarov)

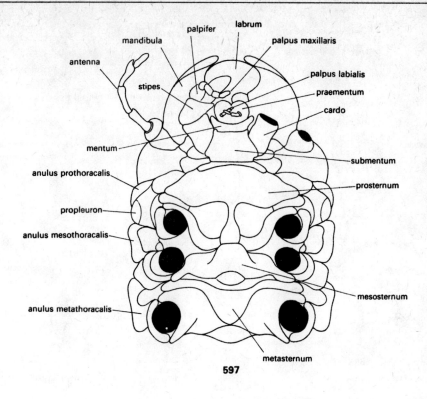

Fig. 597. Head and thorax of Coleoptera (Melolontha-type) in ventral view

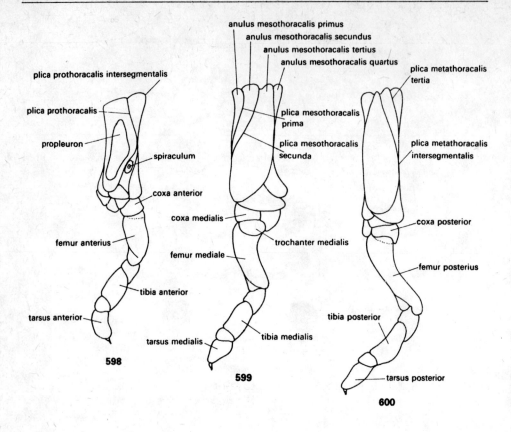

Figs. 598–600. Pro-, meso- and metathoracic segment with leg of Coleoptera in lateral view

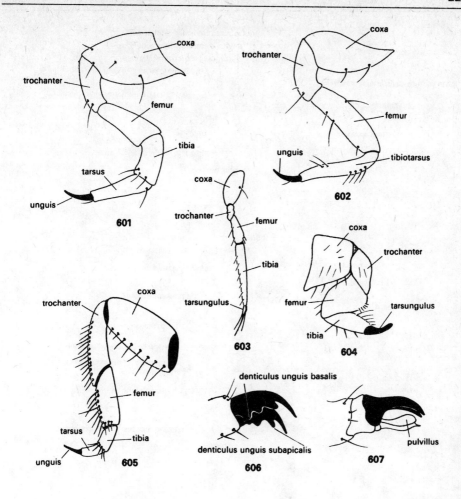

Figs. 601–607. Thoracic leg and parts of Coleoptera. – 601: Typical form, 602: leg with tibiotarsus; 603–605: three types of leg, 603: Dineutes, 604: Eleodes, and 605: Harpalus; 606–607: claw without and with pulvillus (after Meixner 601–602, Peterson 603–604, and Ghilarov 605–607).

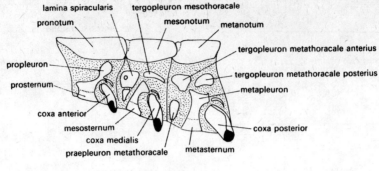

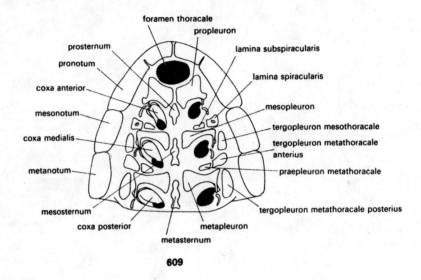

Figs. 608–609. Thorax of Coleoptera (Cassida-type) in lateral and ventral view (after Geisthardt)

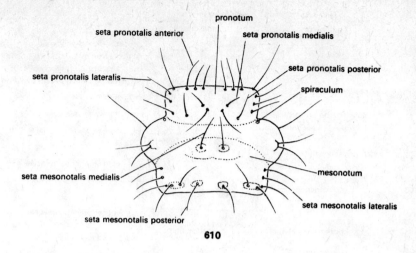

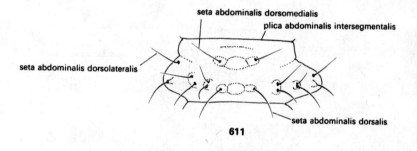

Fig. 610. Pro- and mesothorax of Coleoptera with chaetotaxy in dorsal view. – Fig. 611. Typical abdominal segment with chaetotaxy in dorsal view (after Peterson)

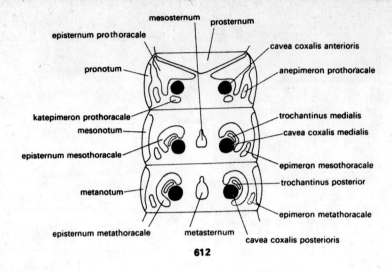

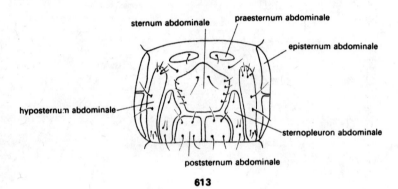

Fig. 612. Schematic drawing of the thorax of Coleoptera in ventral view. – Fig. 613. Typical abdominal segment with sclerites in ventral view (after Ghilarov)

Holomorphosis II: Holometabolia

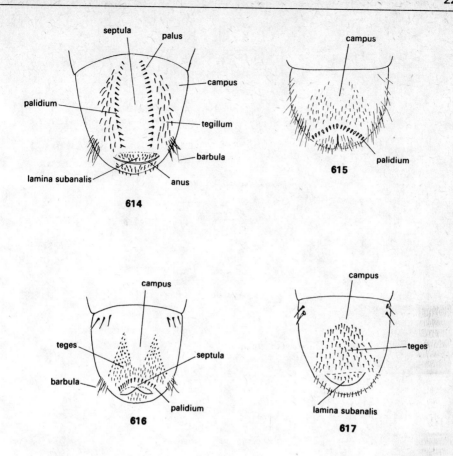

Figs. 614–617. Rasters of Coleoptera. – 614: Hypothetical form, 615–616: two types with palidium, and 617: form without palidium (after Peterson)

15 Morphological

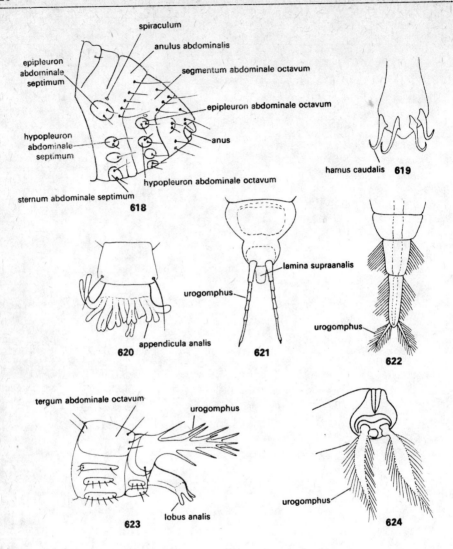

Figs. 618–624. Types of abdominal end of Coleoptera. – 618: Schematic drawing showing sclerites in lateral view, 619: cauda of Dineustes, 620: anal appendages of Photuris, 621: cauda with segmented urogomphus of a Carabid, 622: Dytiscus-type, 623: Ordes-type, and 624: urogomphus of Dytiscus (after Ghilarov 618, Peterson 619–622, and Snodgrass 623–624)

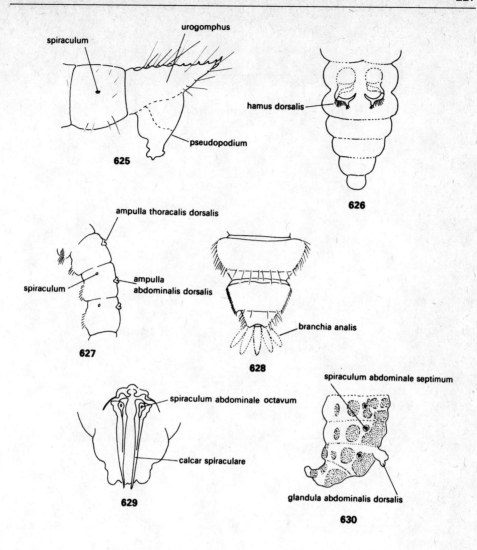

Figs. 625–630. Abdominal parts and special organs of Coleoptera (after Peterson)

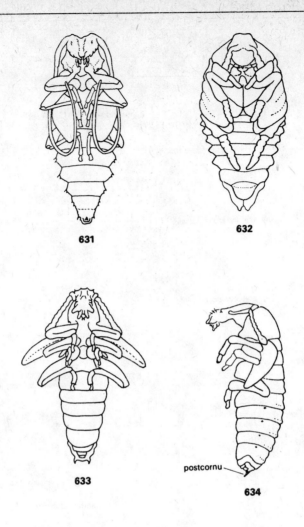

Figs. 631–634. Schematic drawing of some pupa types of Coleoptera. – 631: Cerambyx in ventral, 632: Popillia in ventral, 633–634: Brachyrhynus in ventral and lateral view (after Peterson)

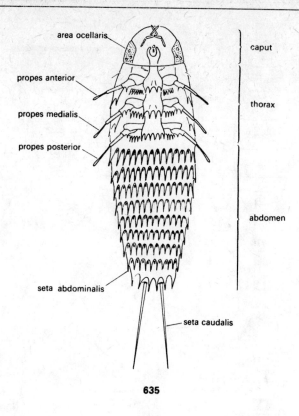

Fig. 635. Schematic drawing of the primary larva of Strepsiptera in ventral view (after Kinzelbach and Kaszab)

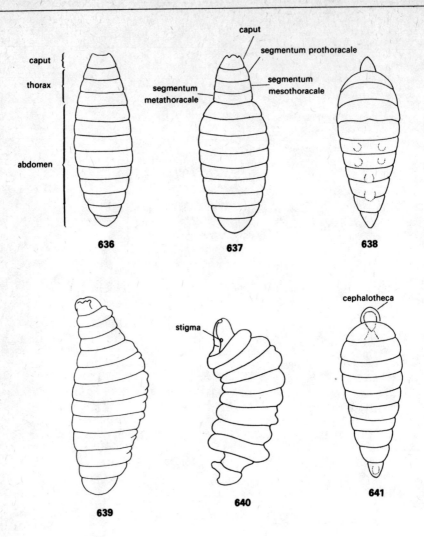

Figs. 636–641. Schematic drawing to show the postembryonic development of female Strepsiptera. – 636: Secondary larva of the 2nd stage in dorsal, 637: same of the 3rd stage in dorsal, 638: tertiary larva in dorsal, 639: secondary larva of the 3rd stage in lateral, 640: tertiary larva in lateral, and 641: puparium in dorsal view (after Kinzelbach)

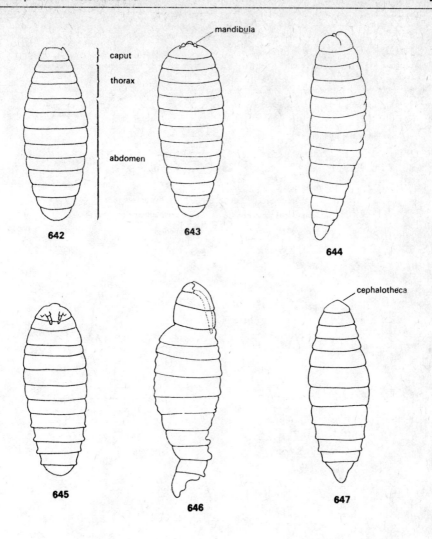

Figs. 642–647. Schematic drawing to show the postembryonic development of male Strepsiptera 642: Secondary larva of the 2nd stage in dorsal, 643–644: same of the 3rd stage in dorsal and lateral, 645–646: tertiary larva in dorsal and lateral, and 647: puparium in dorsal view (after Kinzelbach)

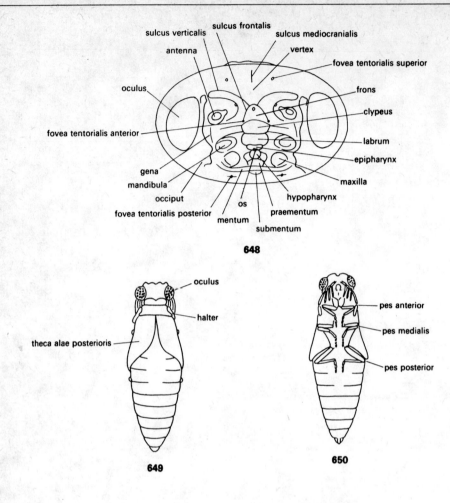

Fig. 648. Schematic drawing of the cephalotheca of Strepsiptera puparium in frontal view. – Figs. 649–650. Pupa in dorsal and ventral view (after Kinzelbach)

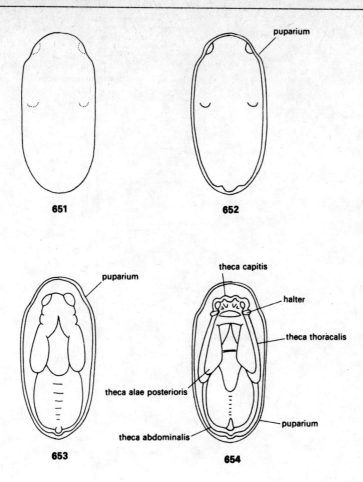

Figs. 651–654. The final stages of the postembryonic development of male Strepsiptera. – 651: Last larval stage, 652: prepupa, 653: pupa, and 654: subimago (after Kinzelbach)

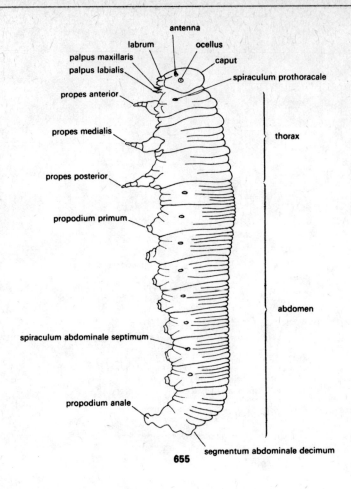

Fig. 655. Schematic drawing of the Hymenoptera in lateral view

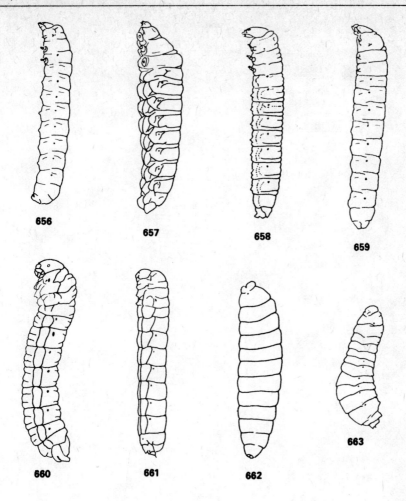

Figs. 656–663. Characteristic types of Hymenoptera larva. – 656: Metallus, 657: Phlebatrophia, 658: Profenusa, 659: Kaliofenusa, 660: Janus, 661: Tremex, 662: Harmolita, and 663: Orussus (after Yuasa 656–657, 659–661, 663, and Peterson 658, 662)

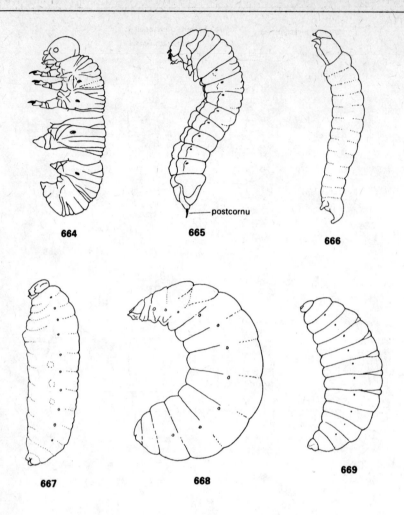

Figs. 664–669. Characteristic types of Hymenoptera larva. – 664: Zaraea, 665: Xiphydria, 666: Macrocentrus, 667: Vespa, 668: Apis, and 669: Torymus (after Peterson)

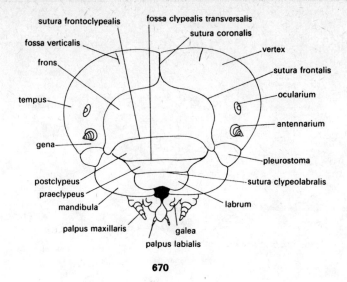
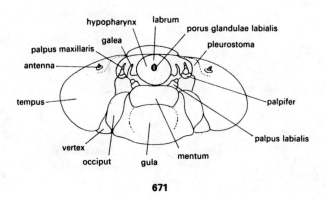

Figs. 670–671. Head of Hymenoptera in dorsal and ventral view (after Yuasa)

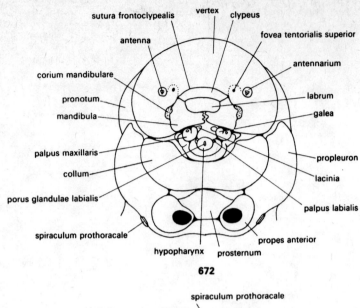
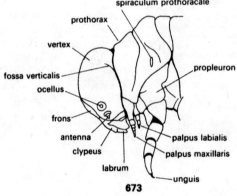

Figs. 672–673. Head and prothorax of Hymenoptera in frontal and lateral view. – 672: Tremex-type, and 673: Dolerus-type (after Yuasa)

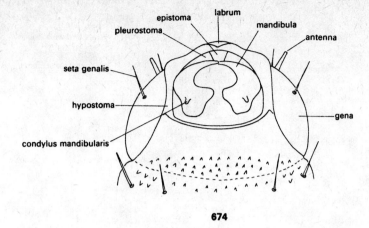

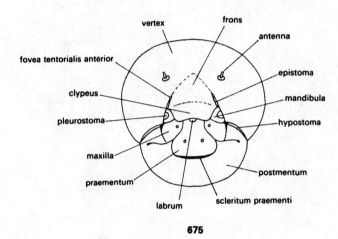

Figs. 674–675. Head of Hymenoptera in ventral and frontal view. – 674: Eurytoma-type, and 675: Sphecophaga-type (after Parker 674, and Snodgrass 675)

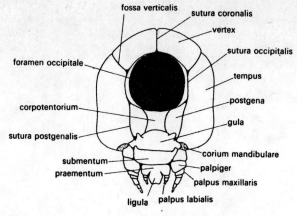

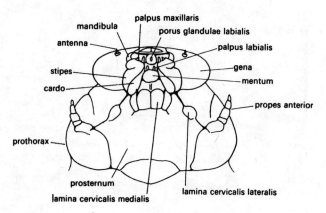

Fig. 676. Head of Hymenoptera in posterior view. – Fig. 677. Head and prothorax in ventral view (after Yuasa)

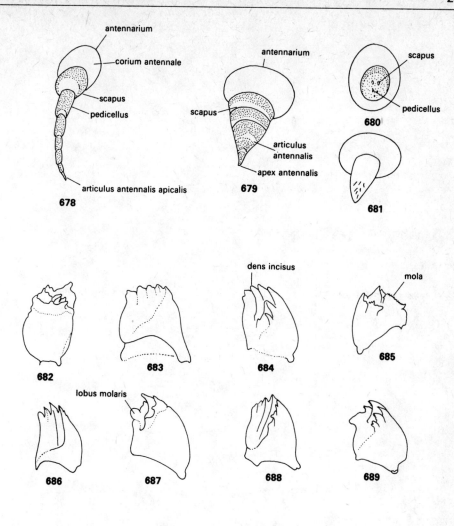

Figs. 678–681. Types of antenna of primitive Hymenoptera. – 678: Neurotoma, 679: Strongylogaster, 680: Cimbex, and 681: Zaraea. – Figs. 682–689. Types of mandible of primitive Hymenoptera. – 682: Cimbex, 683: Diprion, 684: Strongylogaster, 685: Allantus (right), 686: Sofus, 687: Arge, 688: Erythraspides, and 689: Allantus (left) (after Peterson)

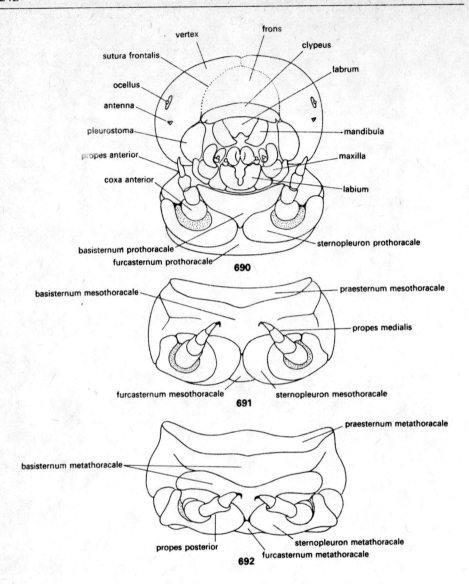

Figs. 690–692. Head with prothorax, meso- and metathorax of Hymenoptera in ventral view

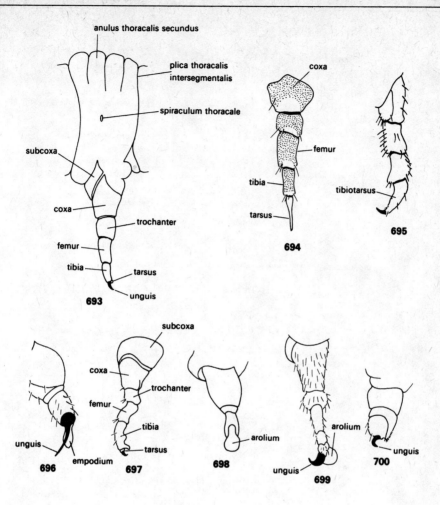

Figs. 693–700. Proleg of Hymenoptera. – 693: Typical form with thoracic segment; 694–700: types in primitive Hymenoptera, 694: Neurotoma, 695: Dolerus, 696: Acordulecera, 697: Megaxyela, 698: Sofus, 699: Arge, and 700: Endelomyia (after Peterson)

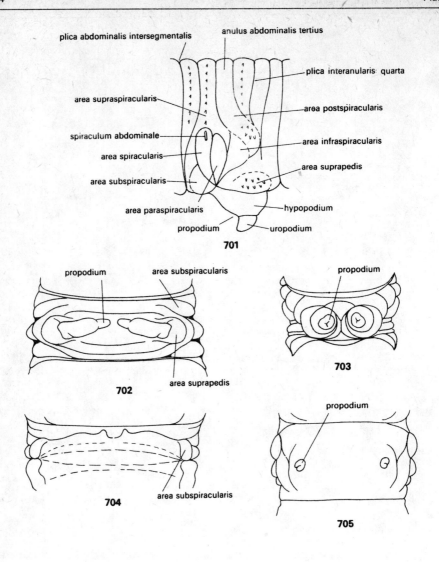

Fig. 701. Theoretical abdominal segment of Hymenoptera in lateral view. – Figs. 702–705. Some types of the third abdominal segment in ventral view. – 702: Megaxyela, 703: Neodiprion, 704: Pamphilius, and 705: Metallus (after Peterson 701, and Yuasa 702–705)

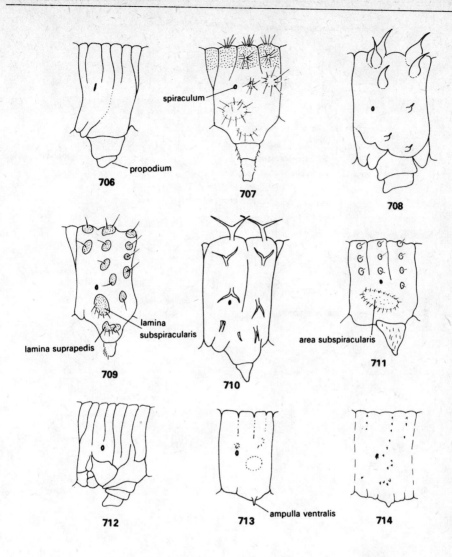

Figs. 706–714. Some types of the third abdominal segment of Hymenoptera in lateral view. – 706: Tomostethus, 707: Priophorus, 708: Erythraspides, 709: Nematus, 710: Monophadnoides, 711: Arge, 712: Neodiprion, 713: Acordulecera, and 714: Sofus (after Peterson)

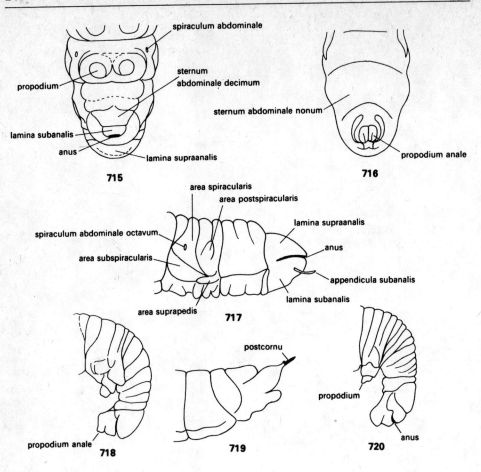

Figs. 715–720. Some types of the abdominal end of Hymenoptera. – 715: Caliroa in ventral, 716: Metallus in ventral, 717: Pamphilius in lateral, 718: Dolerus in lateral, 719: Tremex in lateral, and 720: Neodiprion in lateral view (after Yuasa)

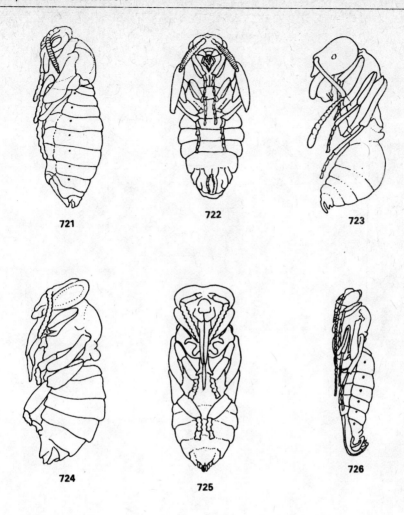

Figs. 721–726. Some types of pupa of Hymenoptera. – 721–722: Neodiprion in lateral and ventral, 723: Formica in lateral, 724–725: Apis in lateral and ventral, and 726: Ichneumonoid form in lateral view (after Peterson)

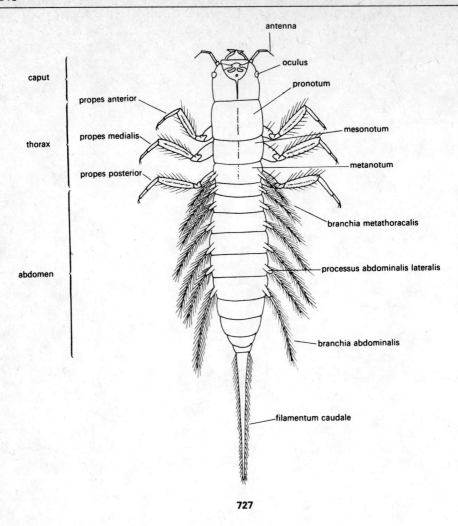

Fig. 727. Schematic drawing of the Megaloptera in dorsal view

Holomorphosis II: Holometabolia

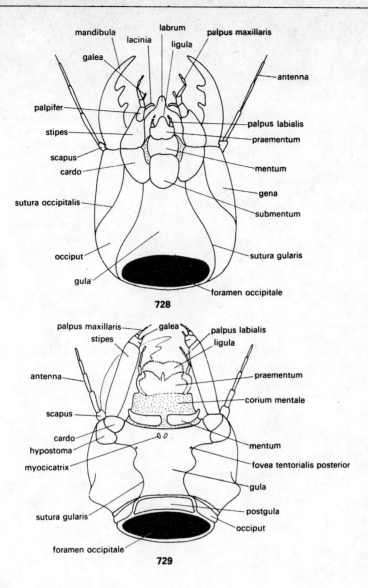

Figs. 728–729. Head of Megaloptera in ventral view. – 728: Sialis-type, and 729: Corydalus-type (after Bertrand 728, and Peterson 729)

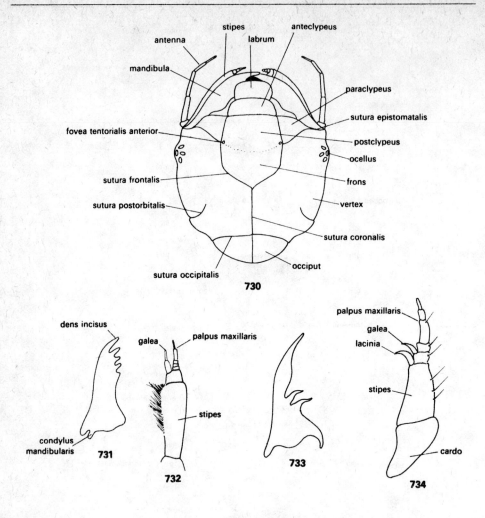

Fig. 730. Head of Megaloptera in dorsal view. – Figs. 731–734. Mouthparts. – 731–732: The mandible and the maxilla of the Chauliodes-type, 733–734: same of the Sialis-type (after Bertrand 730, and Peterson 731–734)

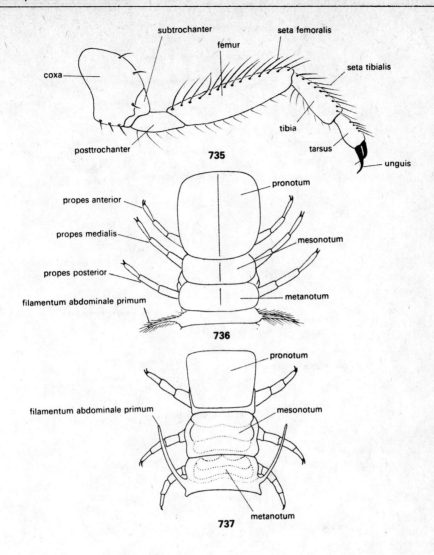

Fig. 735. Typical thoracic leg of Megaloptera. – Figs. 736–737. Thorax and first abdominal segment in dorsal view. – 736: Corydalus-type, and 737: Chauliodes-type (after Bertrand, and Peterson 736–737)

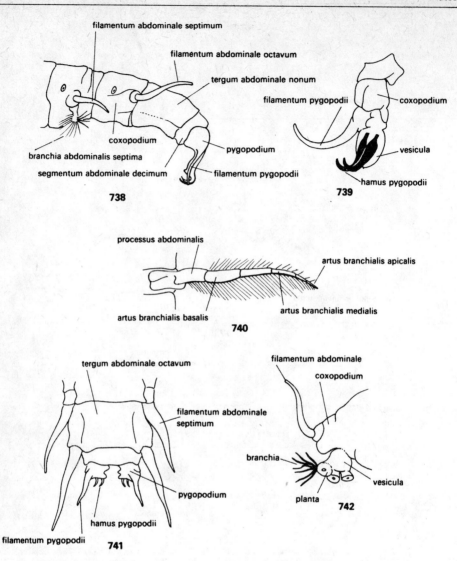

Figs. 738–742. Abdominal end and appendages of Megaloptera. – 738–739: Abdominal end in lateral and pygopodium in ventral view of Corydalus, 740: branchia of Sialis, 741: abdominal end of Chauliodes in dorsal view, and 742: an abdominal appendage of Corydalus in posterior view (after Snodgrass)

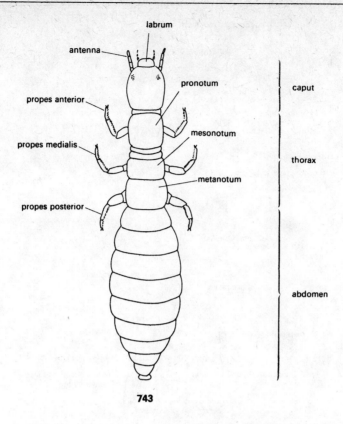

Fig. 743. Schematic drawing of the Raphidioptera in dorsal view (after Steinmann)

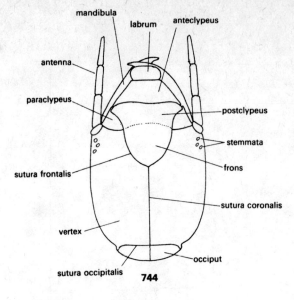

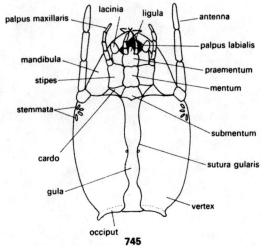

Figs. 744–745. Head of Raphidioptera in dorsal and ventral view (after Crampton)

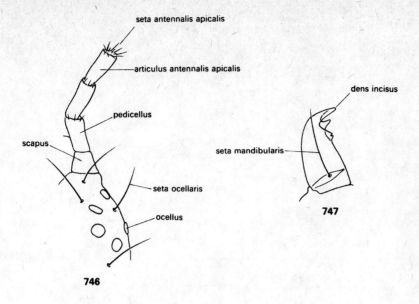
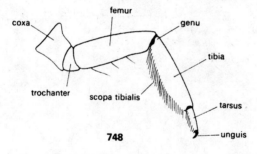

Fig. 746. Antenna with ocellar area of Raphidioptera. − Fig. 747. The mandible. − Fig. 748. Thoracic leg (after Aspöck 746–747)

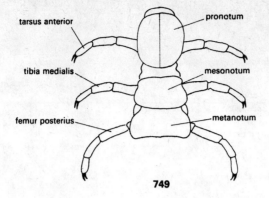

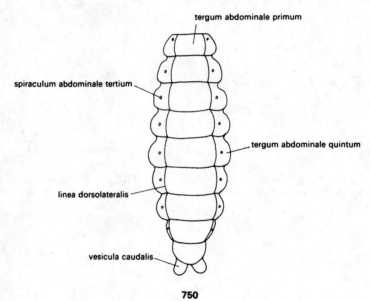

Figs. 749–750. Thorax and abdomen of Raphidioptera in dorsal view (after Peterson)

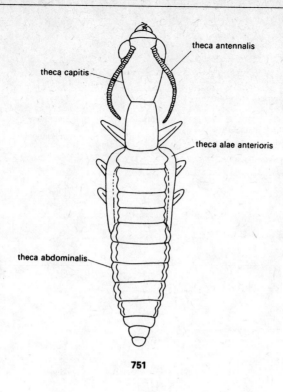

Fig. 751. Pupa of Raphidioptera in dorsal view (after Steinmann)

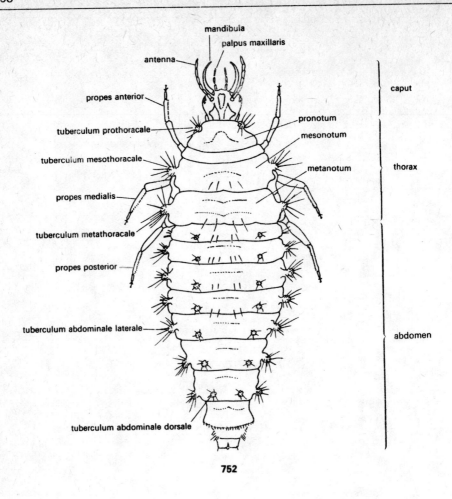

Fig. 752. Schematic drawing of the Neuroptera in dorsal view

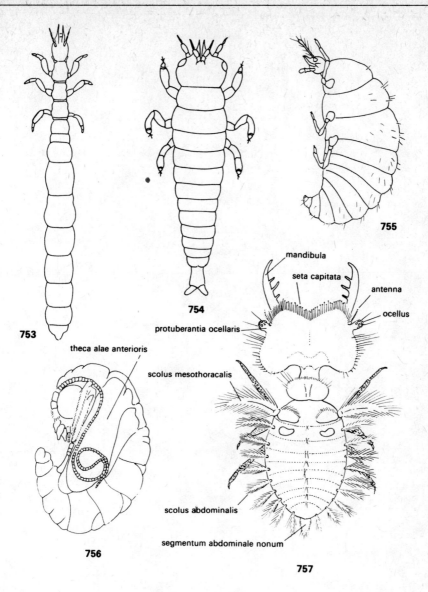

Figs. 753–757. Some types of Neuroptera larva and a typical pupa. – 753: Nallachius larva in dorsal, 754: Mantispa larva in dorsal, 755: Conwentzia larva in lateral, 756: Chrysopa pupa in lateral, and 757: Colobopterus larva in dorsal view (after Peterson)

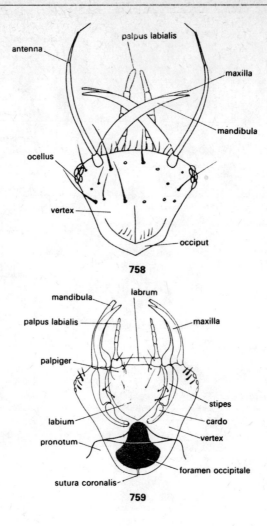

Figs. 758–759. Head of Neuroptera (Chrysopa-type) in dorsal and ventral view (after Rousset 758, and Grassé 759)

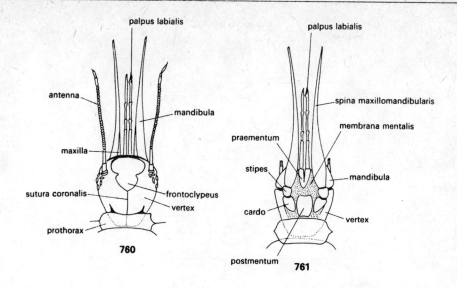

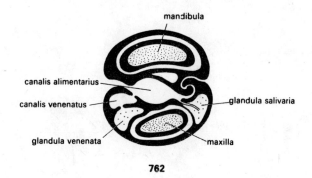

Figs. 760–761. Head of Neuroptera (Osmylus-type) in dorsal and ventral view. – Fig. 762. The maxillomandibular lancet of Osmylus in cross-section (after Wundt)

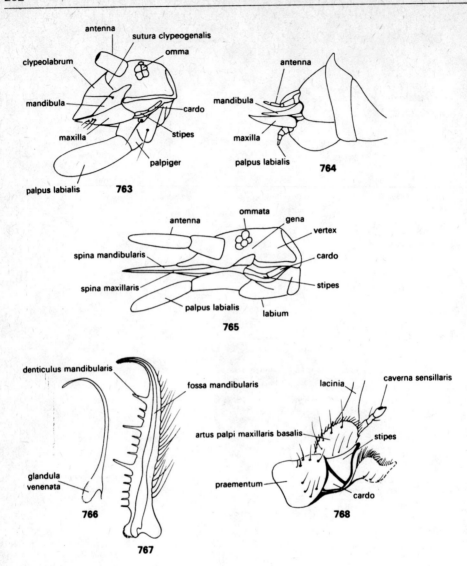

Figs. 763–765. Three types of Neuroptera head in lateral view. – 763: Coniopteryx, 764: Ithone, and 765: Aleuropteryx. – Figs. 766–767. The lacinia and mandible of Myrmeleon. – Fig. 768. The maxilla of Myrmeleon (after Rousset 763, 765, and Grassé 764, 766–768)

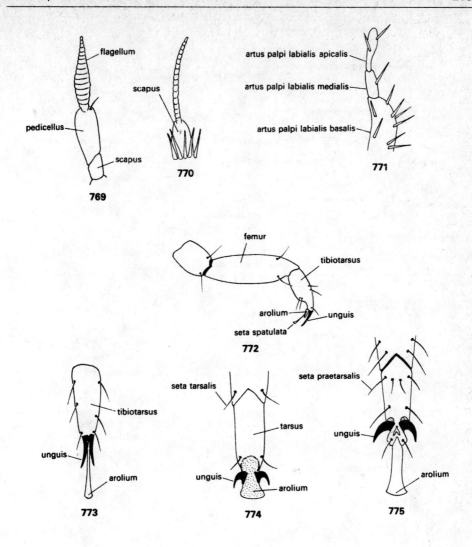

Figs. 769–770. Two types of Neuroptera antenna. – 769: Mantispa, and 770: Ascalaphus. – Fig. 771. The labial palp of Ascalaphus. – Figs. 772–775. Thoracic leg and parts. – 772–773: Fore leg and tibiotarsus of middle leg of Nallachius; 774–775: two types of pretarsus, 774: Hemerobius, and 775: Chrysopa (after Peterson)

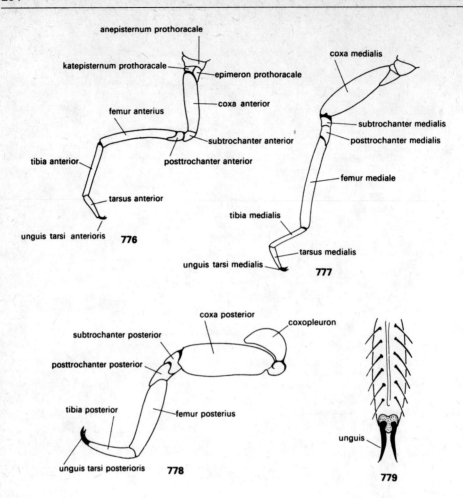

Figs. 776–779. Fore, middle and hind leg, and pretarsus of Neuroptera (Myrmeleon-type)

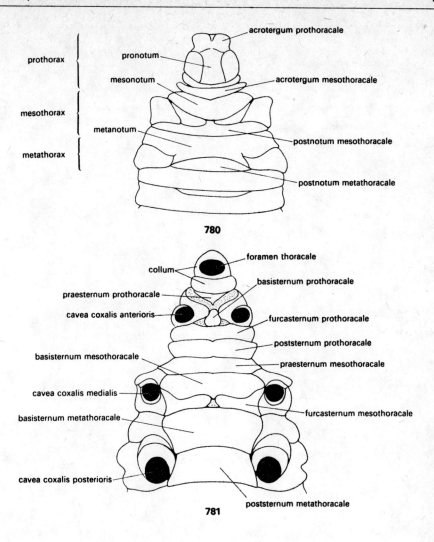

Figs. 780–781. Thorax of Neuroptera (Myrmeleon-type) in dorsal and ventral view

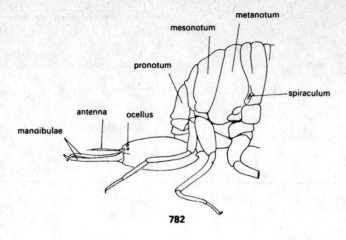

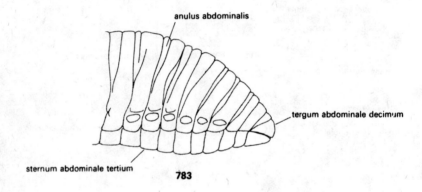

Figs. 782-783. Head and thorax, and abdomen of Neuroptera (Myrmeleon-type) in lateral view

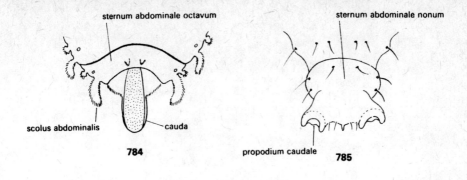

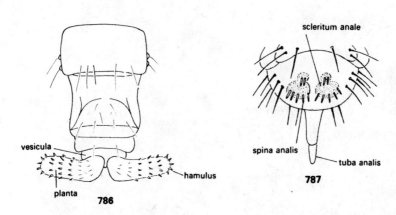

Figs. 784–787. Abdominal end of Neuroptera. – 784: Ascalaphus-type in ventral, 785: Nallachia-type in ventral, 786: Osmylus-type in dorsal, and 787: Euroleon-type in dorsal view (after Peterson 784–785, and Grassé 786–787)

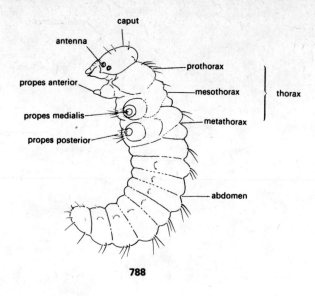
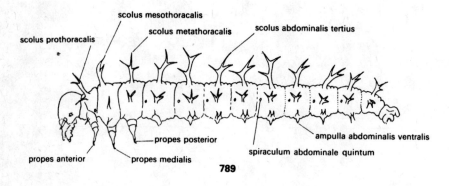

Figs. 788–789. Schematic drawing of the Mecoptera in lateral view. – 788: Boreus-type, and 789: Bittacus-type (after Peterson)

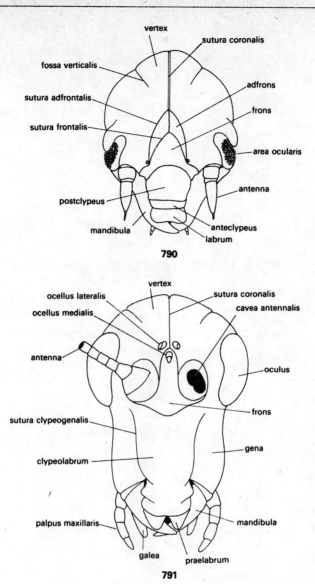

Figs. 790–791. Head of Mecoptera (Panorpa-type) in frontal view. – 790: Larva, and 791: pupa (after Matsuda)

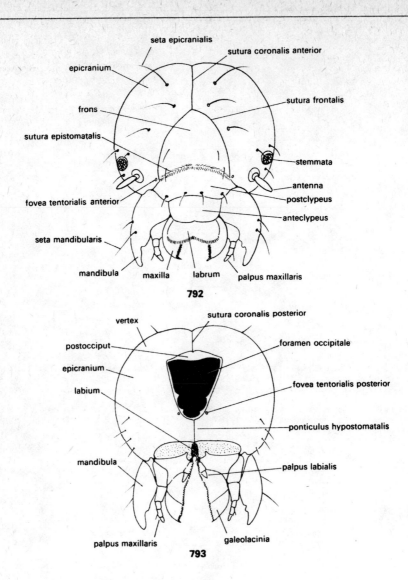

Figs. 792-793. Head of Mecoptera (Bittacus-type) in frontal and ventral view (after Grassé)

Holomorphosis II: Holometabolia

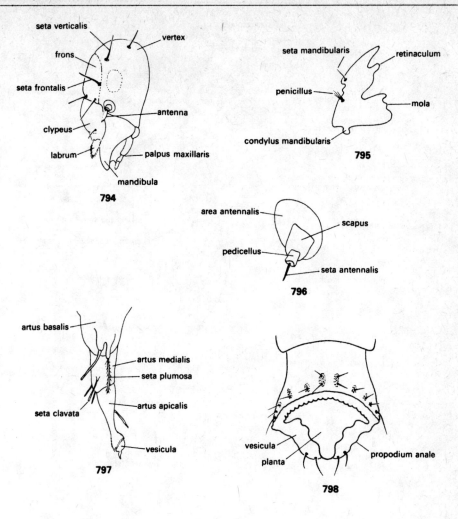

Fig. 794. Head of Mecoptera (Panorpa-type) in lateral view. – Fig. 795. The mandible. – Fig. 796. The antenna. – Fig. 797. The hind leg of Panorpa. – Fig. 798. The abdominal end of Boreus (after Peterson)

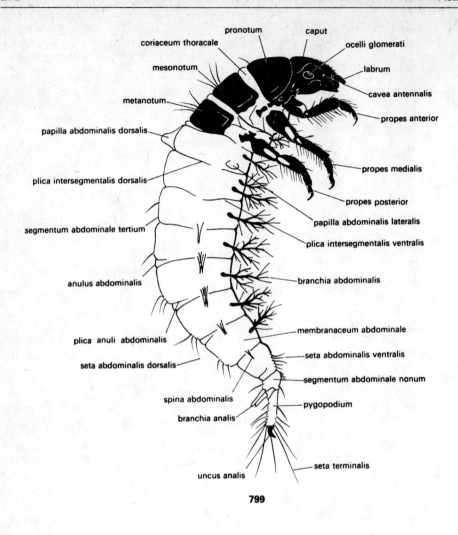

Fig. 799. Schematic drawing of the Trichoptera in lateral view (after Grassé)

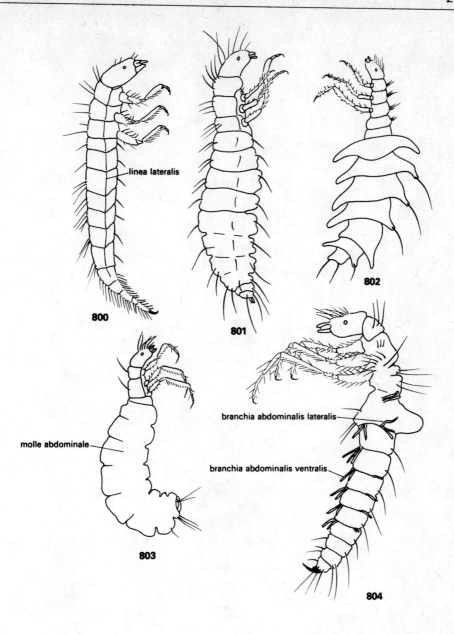

Figs. 800–804. Some types of Trichoptera larva in lateral view. – 800: Cyrnus, 801: Agraylea, 802: Ithytricha, 803: Oxyethira, and 804: Molanna (after Steinmann)

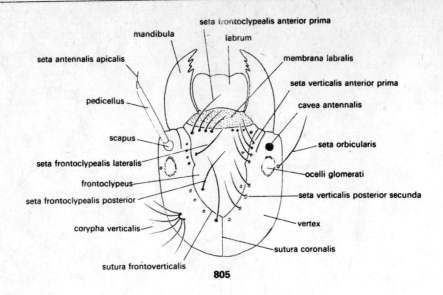

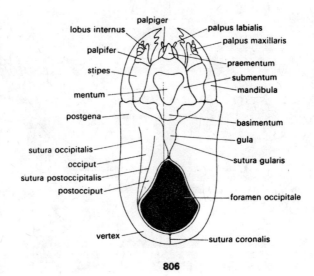

Figs. 805–806. Head of Trichoptera in dorsal and ventral view

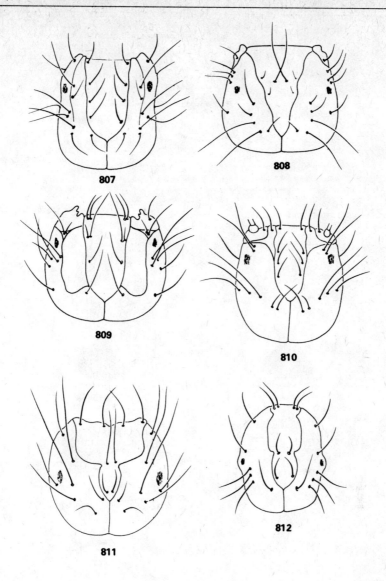

Figs. 807–812. Principal types of Trichoptera head in dorsal view. – 807: Odontocerus, 808: Oecetis, 809: Athripsodes, 810: Trianodes, 811: Limnephilus, and 812: Lepidostoma (after Steinmann)

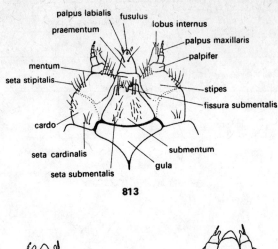
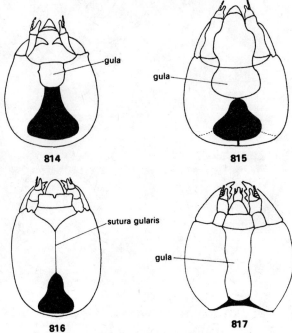

Figs. 813–817. Principal types of Trichoptera head in ventral view. – 813: Mouthparts of Hydropsyche, 814: Leptocerus, 815: Limnephilus, 816: Hydropsyche, and 817: Brachycentrus (after Malicky)

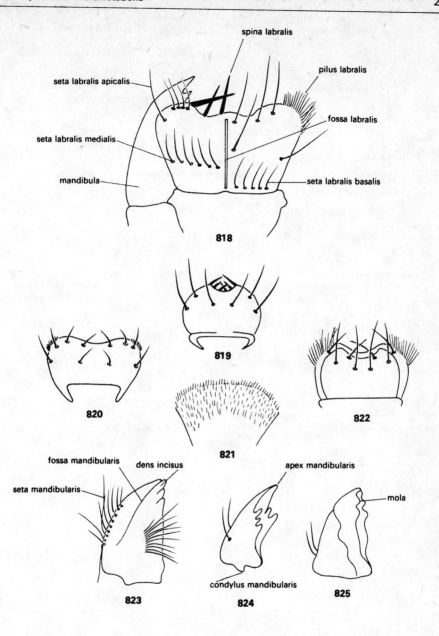

Figs. 818–825. Mouthparts of Trichoptera. – 818–822: The labrum, 818: typical form, 819: Oligostoma, 820: Phryganea, 821: Hydropsyche, and 822: Limnephilus; 823–825: three types of mandible, 823: Cheumatopsyche, 824: Oligostoma, and 825: Athripsodes (after Bertrand 821, Steinmann 822, 824–825)

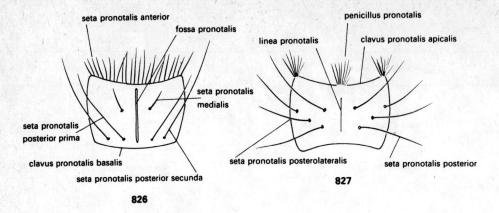
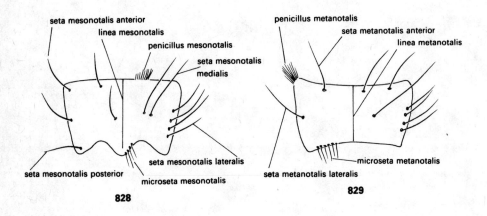

Figs. 826–829. Chaetotaxy of pro-, meso- and metanotum of Trichoptera

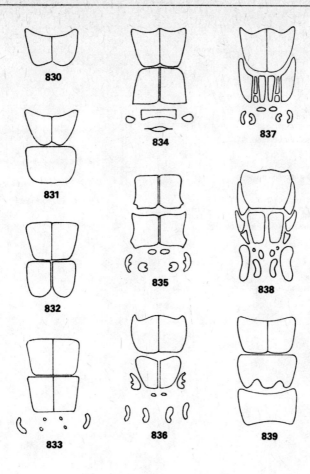

Figs. 830–839. Sclerotization of the thoracic notum of Trichoptera. – 830: Phryganea. 831: Sericostoma, 832: Beraeodes, 833: Lepidostoma, 834: Odontocerus, 835: Limnephilus, 836: Goera, 837: Silo, 838: Silo, and 839: Hydropsyche (after Bertrand)

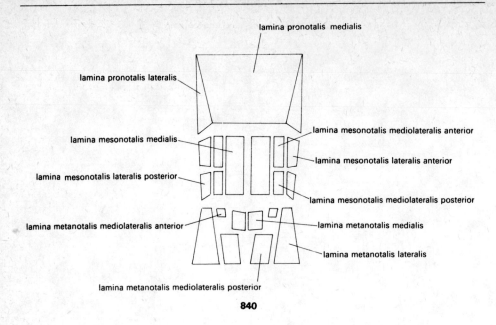

Fig. 840. Diagrammatic representation of the sclerites of the thoracic notum in Trichoptera

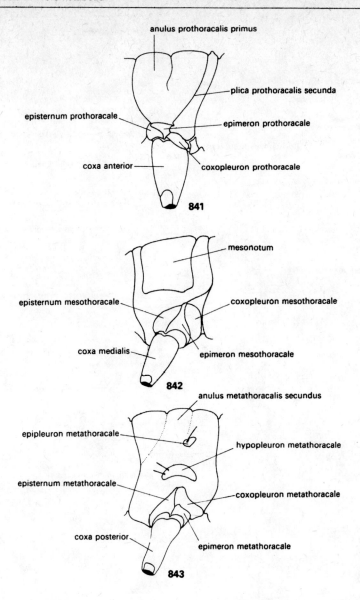

Figs. 841–843. Pro-, meso- and metathorax with coxa of Trichoptera in lateral view

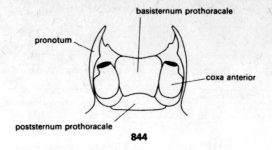

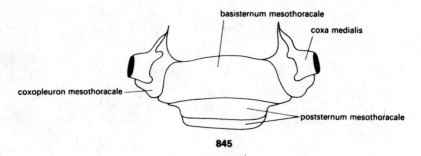

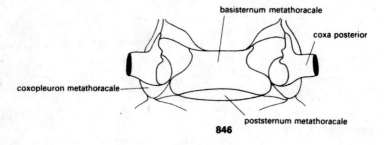

Figs. 844–846. Pro-, meso- and metathorax with coxa of Trichoptera in ventral view

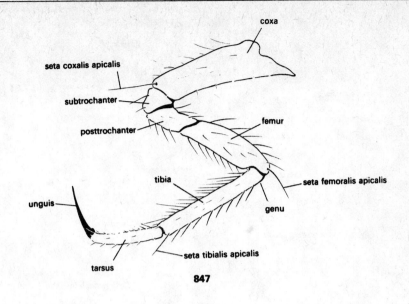
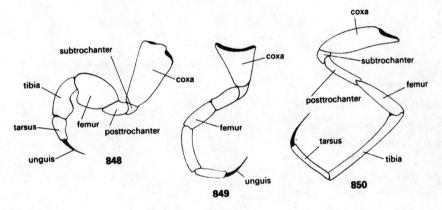

Figs. 847–850. Types of thoracic leg of Trichoptera. – 847: General, 848: fore, 849: middle, and 850: hind

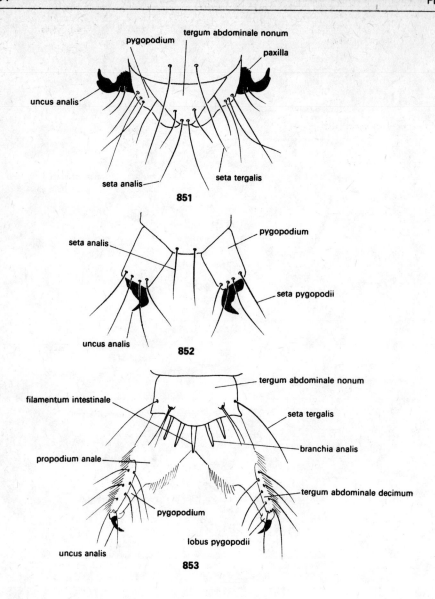

Figs. 851–853. Abdominal end of Trichoptera in dorsal view. – 851: Neuronia-type, 852: Phryganea-type, and 853: Plectrocnemia-type (after Steinmann 851–852, and Grassé 853)

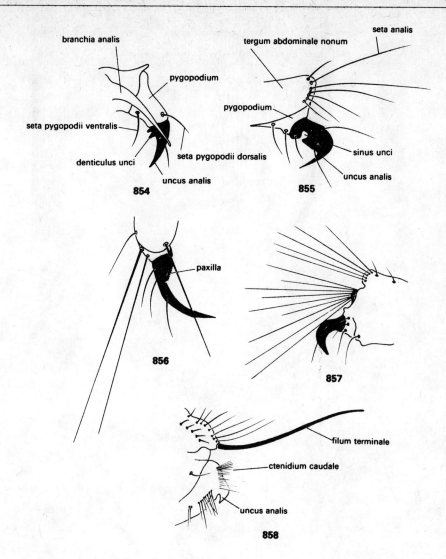

Figs. 854–858. Some types of anal hook of Trichoptera. – 854: Rhyacophila, 855: Hydropsyche, 856: Agraylea, 857: Cheumatopsyche, and 858: Beraemyia (after Steinmann)

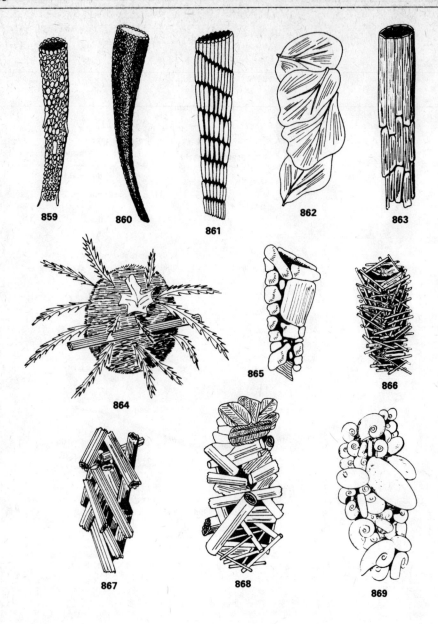

Figs. 859–869. Some types of Trichoptera larva cases. – 859: Mystacides, 860: Athripsodes, 861: Trianodes, 862: Grammotaulius, 863: Macrotaulius, and 864–869: Limnephilus spp. (after Steinmann)

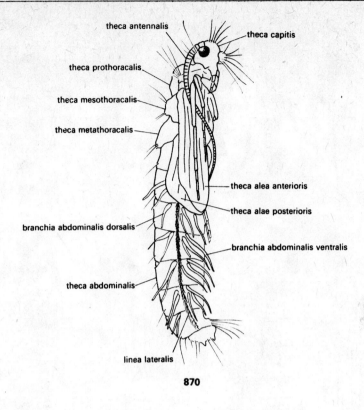

Fig. 870. Schematic drawing of Trichoptera pupa (after Steinmann)

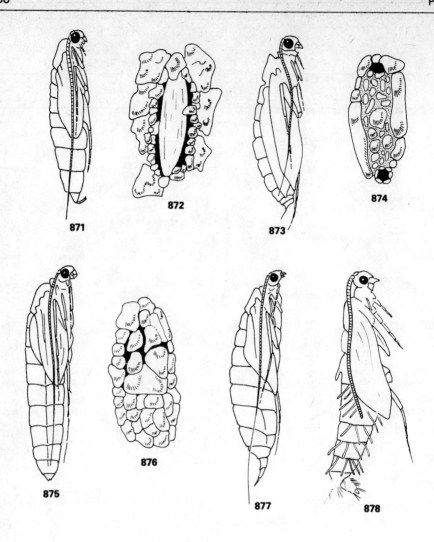

Figs. 871–878. Some types of pupa and cases of Trichoptera. – 871–872: Glossosoma, 873–874: Agapethus, 875–876: Polycentropus, 877: Hydropsyche pupa, and 878: Neuronia pupa (after Steinmann 871–877, and Bertrand 878)

Holomorphosis II: Holometabolia

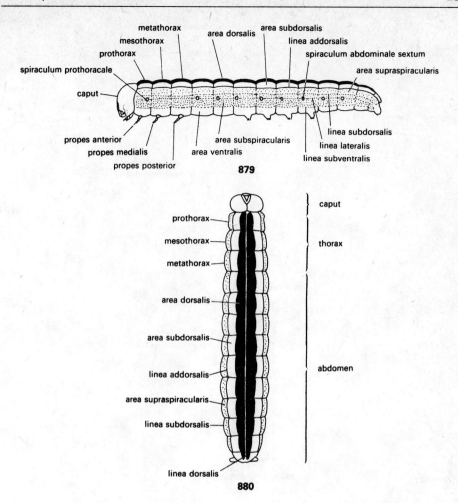

Figs. 879–880: Schematic drawing to show the several longitudinal coloration areas of Lepidoptera in lateral and dorsal view (after Peterson)

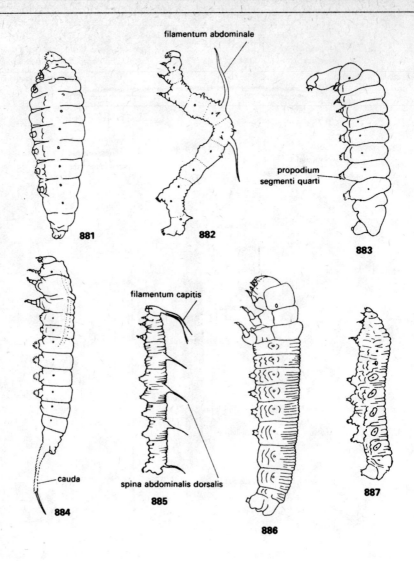

Figs. 881–887. Characteristic types of Lepidoptera larva in lateral view. – 881: Carpocapsa, 882: Nematocampa, 883: Lycaenid, 884: Cerura, 885: Athena, 886: Thyridopteryx, and 887: Pandorus (after Peterson)

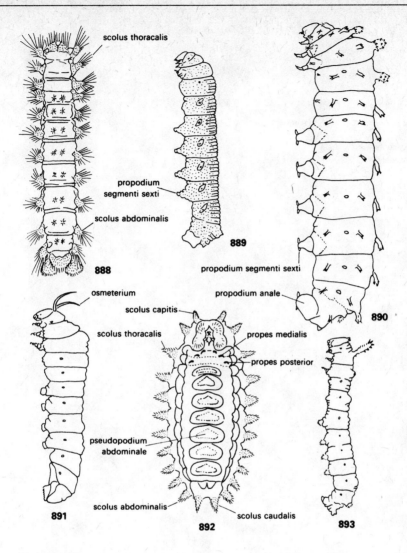

Figs. 888–893. Characteristic types of Lepidoptera larva. – 888: Tolype in dorsal, 889: Pholus in lateral, 890: Samia in lateral, 891: Papilio in lateral, 892: Euclea in ventral and 893: Basilarchia in lateral view (after Peterson)

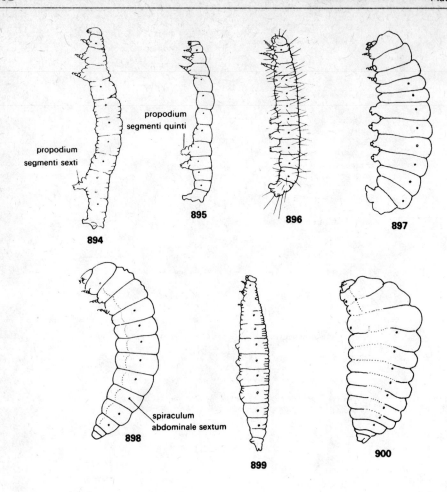

Figs. 894–900. Characteristic types of Lepidoptera larva in lateral view. – 894: Coryphista, 895: Trichoplusia, 896: Scolecocampa, 897: Strymon, 898: Tegeticula, 899: Aesiale, and 900: Prodoxus (after Peterson)

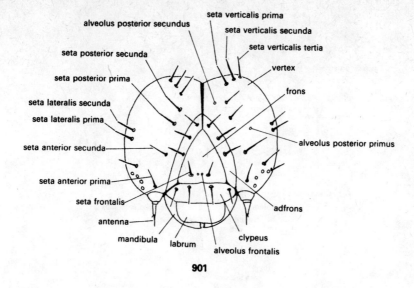
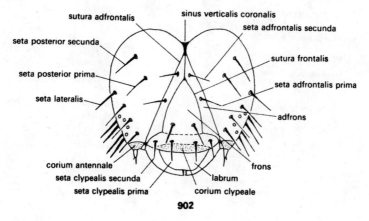

Figs. 901–902. Two types of Lepidoptera head with chaetotaxy (after Peterson)

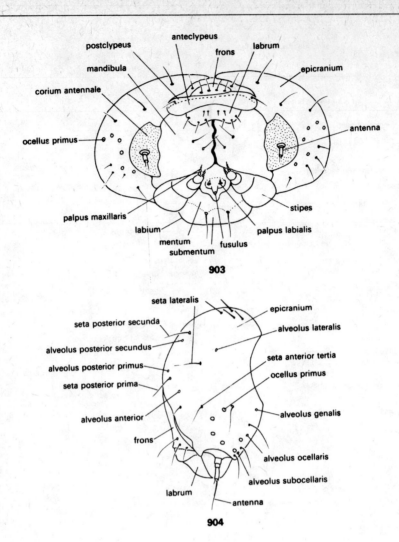

Figs. 903–904. Head of Lepidoptera in ventral and lateral view (after Peterson 903, and Beck 904)

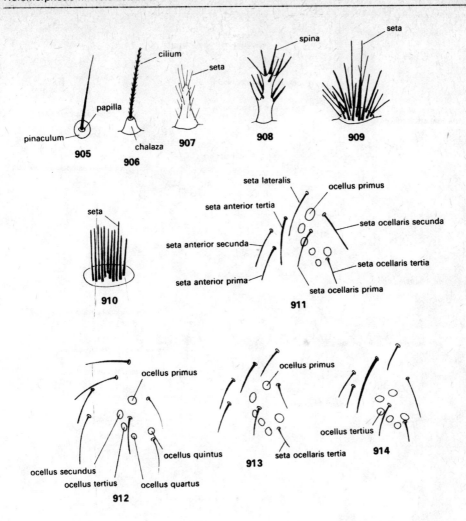

Figs. 905–910. Tegumentary formations of Lepidoptera. – 905: Seta, 906: chalaza with ciliate seta, 907: sentus, 908: scolus, 909: verruca, and 910: verricule. – Figs. 911–914. Some chaetotaxy of the ocellar area. – 911: Gnorimoschema, 912: Dichomeris, 913: Grapholitha, and 914: Archips (after Peterson)

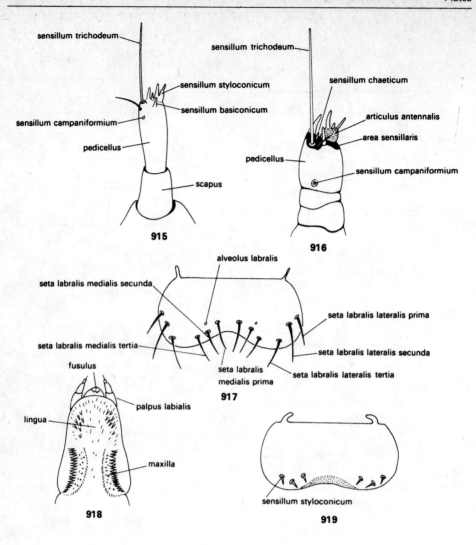

Figs. 915–916. Two types of antenna of Lepidoptera. – 915: Typical, and 916: Cryphia. – Fig. 917: The labrum. – Fig. 918. The hypopharynx. – Fig. 919. The epipharynx (after Peterson 915, 917–919, and Beck 916)

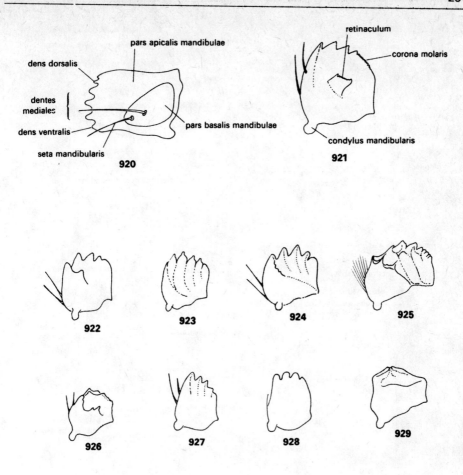

Figs. 920–929. Types of mandible of Lepidoptera. – 920: Typical, 921: Alypia, 922: Ancylis, 923: Laphygma, 924: Platynota, 925: Malacosoma, 926: Nygmia, 927: Prionoxystus, 928: Sitotroga, and 929: Cirphis (after Beck 920, and Peterson 921–929)

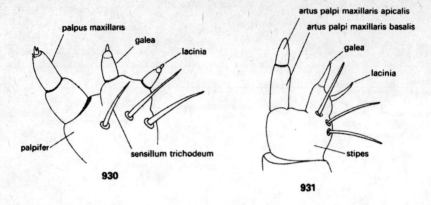
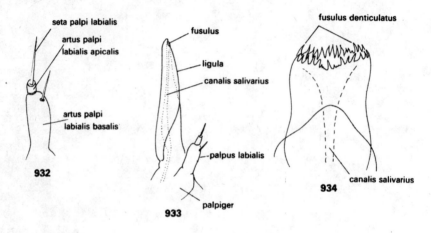

Figs. 930–934. Mouthparts of Lepidoptera. – 930–931: Two types of maxilla, 930: Hypena, and 931: Abrostola; 932: labial palp, 933: half labium with the salivary canal, and 934: end of salivary canal (after Beck)

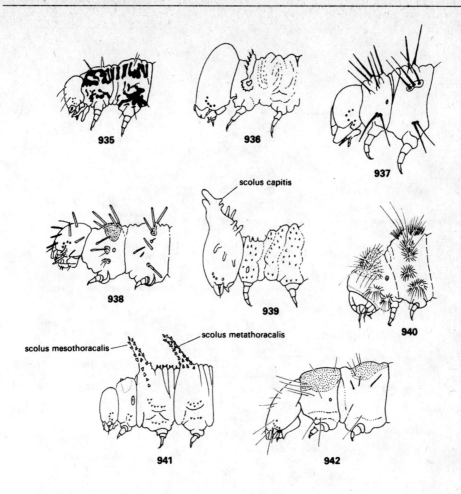

Figs. 935–942. Head and thorax of Lepidoptera in lateral view. – 935: Ceramica-type, 936: Hylephila-type, 937: Hypoprepia-type, 938: Epizeuxis-type, 939: Asterocampa-type, 940: Harrisina-type, 941: Ceratomia-type, and 942: Paraclemensia-type (after Peterson)

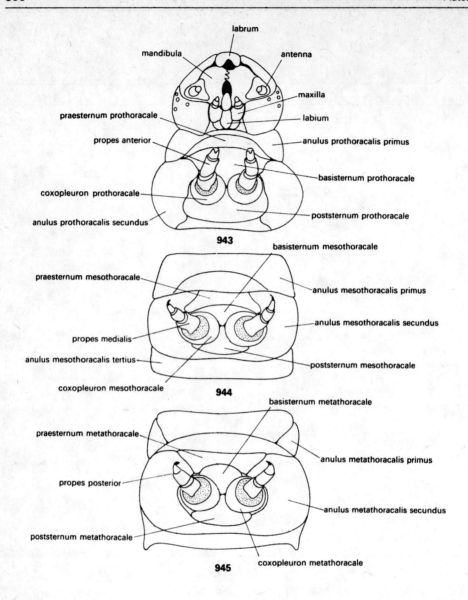

Figs. 943–945. Head and thorax of Lepidoptera (Lasiocampa-type) in ventral view

Holomorphosis II: Holometabolia

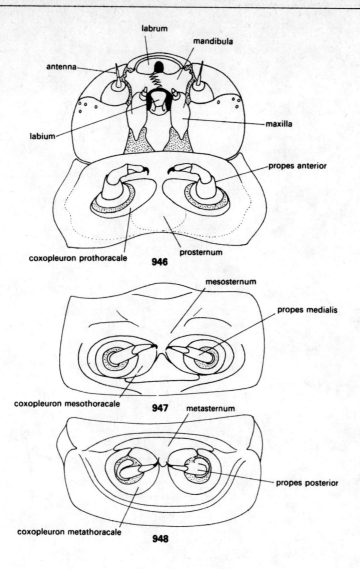

Figs. 946–948. Head and thorax of Lepidoptera (Biston-type) in ventral view

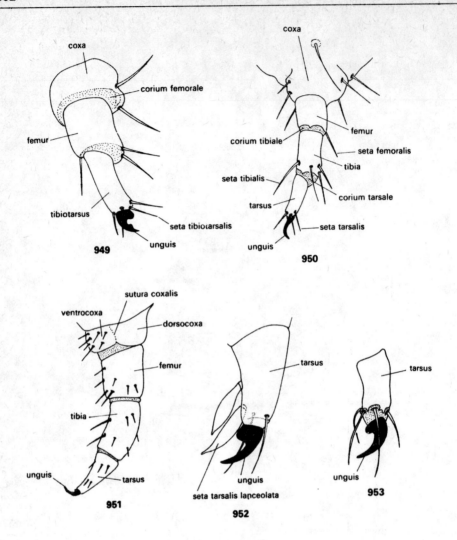

Figs. 949–953. Leg and parts of Lepidoptera. – 949: Autographa, 950: Paradrina, 951: Drepana, 952: Jaspidia, and 953: Papaipema (after Beck 949–952, and Peterson 953)

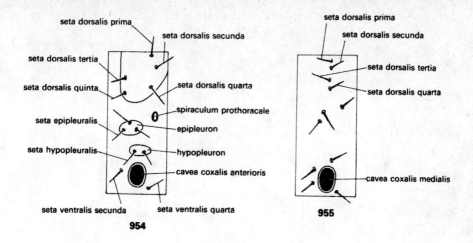
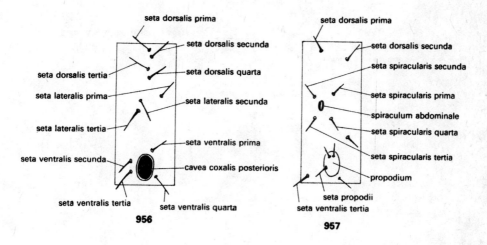

Figs. 954–957. Diagrammatic representation of the thoracic and abdominal chaetotaxy of Lepidoptera in lateral view. – 954: Prothorax, 955: mesothorax, 956: metathorax, and 957: a typical abdominal segment (after Beck)

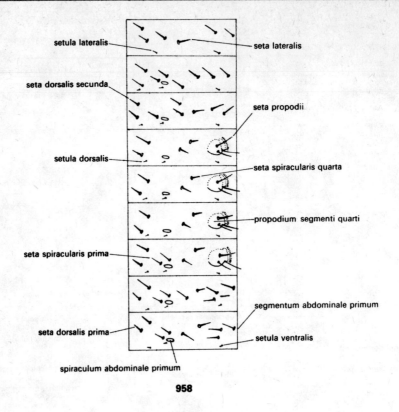

Fig. 958. Diagrammatic representation of the abdomen of Lepidoptera in lateral view (after Beck and Peterson, combined)

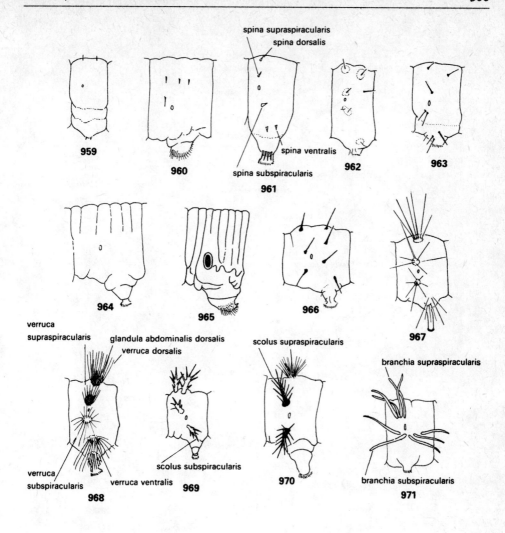

Figs. 959–971. Some types of the fourth abdominal segment of Lepidoptera in lateral view. – 959: Sitotroga, 960: Proteides, 961: Anisota, 962: Ecdytolopha, 963: Prionoxystus, 964: Colias, 965: Protoparce, 966: Alabama, 967: Pterophorus, 968: Nygmia, 969: Polygonia, 970: Hemileuca, and 971: Catoclysta (after Peterson)

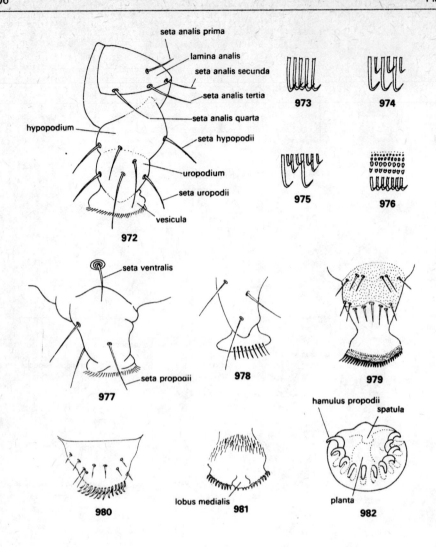

Fig. 972. Typical propodium of Lepidoptera. – Figs. 973–976. Four types of arrangement of crochets on the propodium, 973: uniserial and uniordinal, 974: uniserial and biordinal, 975: uniserial and triordinal, and 976: multiserial and uniordinal. – Figs. 977–981. The propodium of various types. – Fig. 982. The plantar surface of Euxoa (after Beck)

Holomorphosis II: Holometabolia

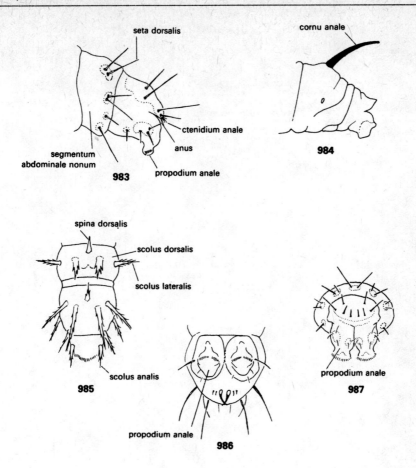

Figs. 983–987. Abdominal end of Lepidoptera. – 983: Carpocapsa in lateral, 984: Protoparce in lateral, 985: Nymphalis in dorsal, 986: Recurvaria in ventral, and 987: Carpocapsa in posterior view (after Peterson)

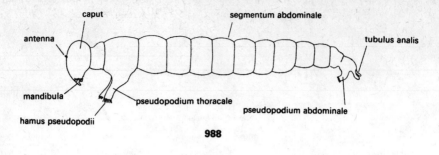

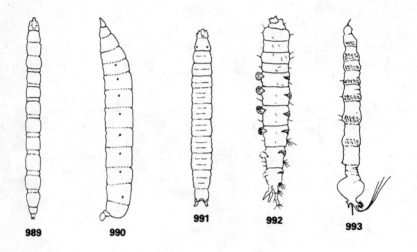

Fig. 988. Schematic drawing of the Diptera (Nematocera-type) in lateral view. – Figs. 989–993. Characteristic types of Diptera larva. – 989: Sylvicola in dorsal, 990: Fungivorid in lateral, 991: Trichocera in dorsal, 992: Antocha in lateral, and 993: Eriocera in lateral view (after Wéber 988, and Peterson 989–993)

Holomorphosis II: Holometabolia

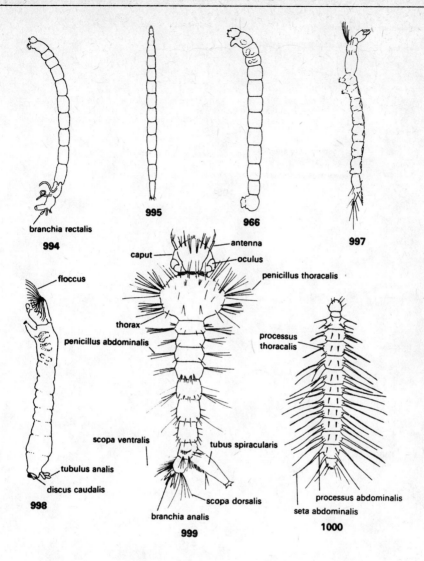

Figs. 994–1000. Characteristic types of Diptera (Nematocera-type) larva. – 994: Chironomus in lateral, 995: Culicoides in dorsal, 996: Spaniotoma in lateral, 997: Dixa in lateral, 998: Simulium in lateral, 999: Aedes in dorsal, and 1000: Atrichopogon in dorsal view (after Peterson)

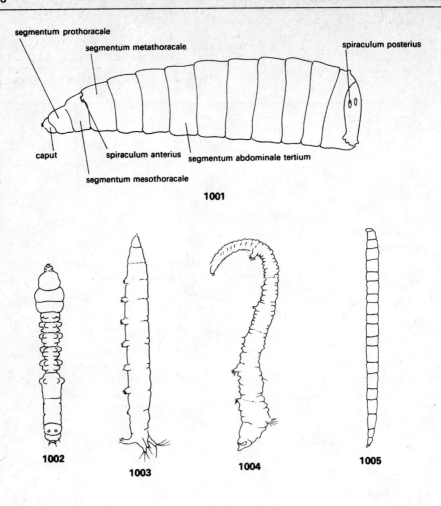

Fig. 1001. Schematic drawing of the Diptera (Brachycera-type) in lateral view. – Figs. 1002–1005. Characteristic types of Diptera larva. – 1002: Laphria in dorsal, 1003: Rhagionid in lateral, 1004: Vermileo in lateral, and 1005: Psilocephala in lateral view (after Peterson)

Holomorphosis II: Holometabolia

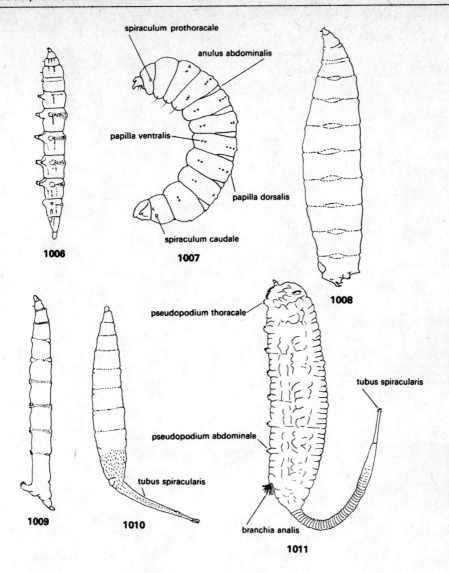

Figs. 1006–1011. Characteristic types of Diptera (Brachycera-type) larva. – 1006: Tabanus in lateral, 1007: Bombyliid in lateral, 1008: Hylemya in lateral, 1009: Limnophora in lateral, 1010: Ephydra in lateral, and 1011: Eristalis in lateral view (after Peterson)

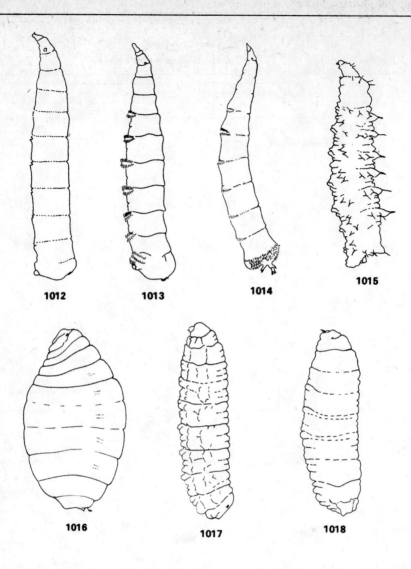

Figs. 1012–1018. Characteristic types of Diptera (Brachycera-type) larva. – 1012: Musca in lateral, 1013: Siphona in lateral, 1014: Sepsis in lateral, 1015: Boccha in lateral, 1016: Eurosta in lateral, 1017: Sturmia in lateral, and 1018: Compsilura in lateral view (after Peterson)

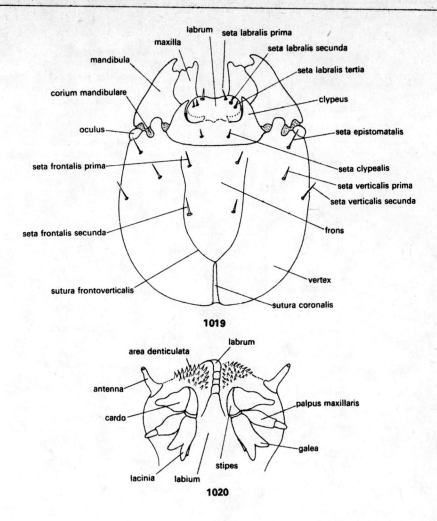

Figs. 1019–1020. Head of Diptera. – 1019: Hesperinus-type in dorsal, and 1020: Rhagio-type in ventral view (after Krivosheina)

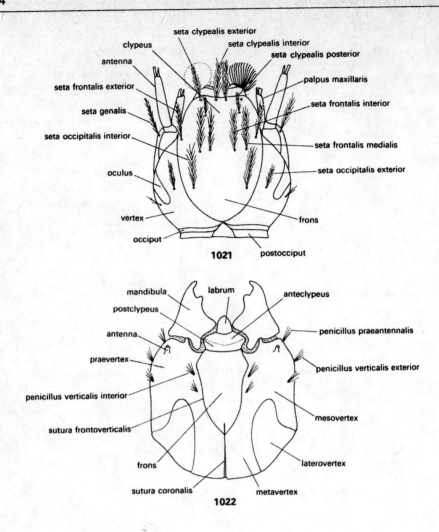

Figs. 1021–1022. Head of Diptera in dorsal view. – 1021: Anopheles-type, and 1022: Protaxymya-type (after Hennig 1021 and Krivosheina 1022)

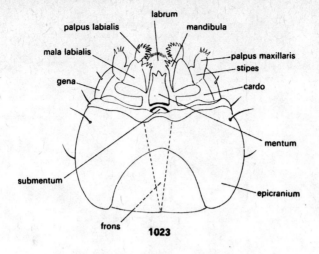

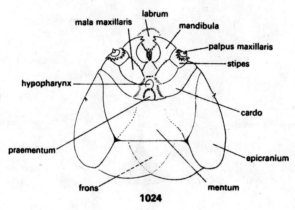

Figs. 1023–1024. Head of Diptera in ventral view. – 1023: Bibio-type, and 1024: Petaurista-type (after Hendel)

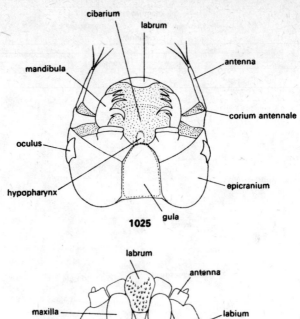
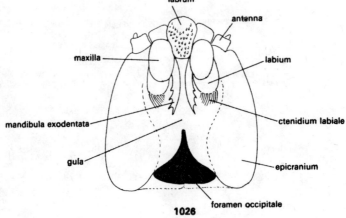

Figs. 1025–1026. Head of Diptera in ventral view. – 1025: Eucorethra-type, and 1026: Mycetobia-type (after Peterson 1025. and Hendel 1026)

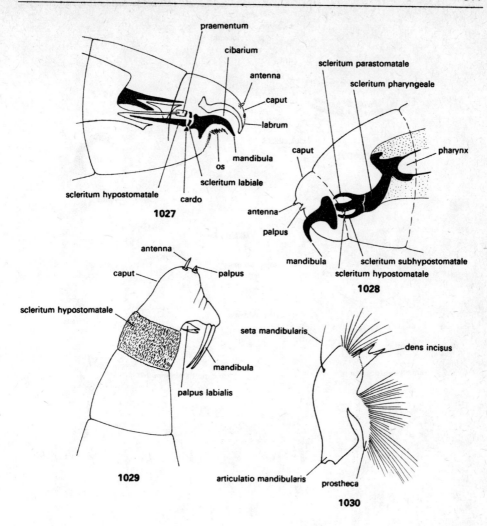

Figs. 1027–1028. Head of Diptera to show internal sclerotization in lateral view. – Fig. 1029. Head of the Melanocheila-type. – Fig. 1030. The mandible of Simulium (after Hendel 1027, Peterson 1028, 1030, and Grassé 1029)

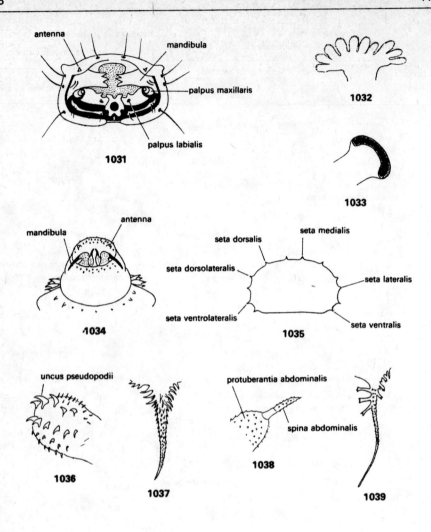

Fig. 1031. Head of Diptera (Bibio-type) in frontal view. – Figs. 1032–1033. Two types of prothoracic spiracle, 1032: Hylemya, and 1033: Palpomyia. – Fig. 1034. Head of the Gasterophilus-type in ventral view. – Fig. 1035. Diagrammatic representation of the abdominal chaetotaxy in cross-section. – Figs. 1036–1039. Some types of body appendages. – 1036: Fourth propodium in lateral view, 1037: enlarged caudal filament, 1038: protuberance with spine, and 1039: normal caudal filament (after Peterson)

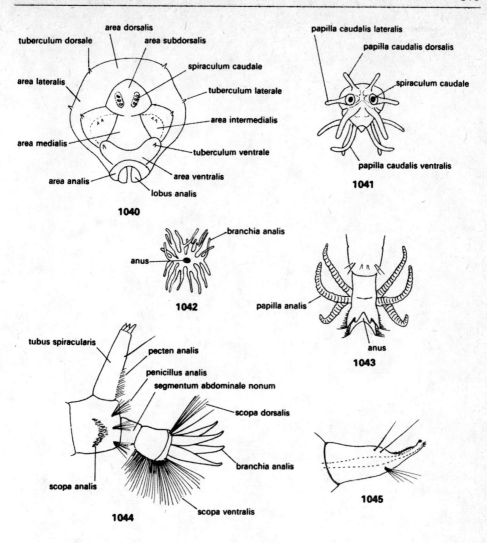

Figs. 1040–1045. Abdominal end of Diptera. – 1040: areas of the typical form, 1041: Prionocera in posterior view, 1042: anal gills, 1043: Tipula in dorsal, 1044: Psorophora in lateral, and 1045: Mansonia sipho in lateral view (after Peterson)

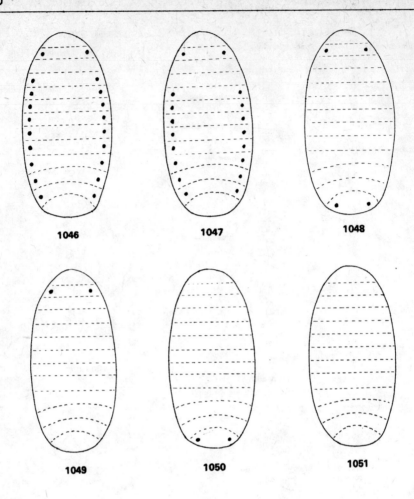

Figs. 1046–1051. The different types of the arrangement of stigmata of Diptera. – 1046: Holopneustic, 1047: peripneustic, 1048: amphipneustic, 1049: propneustic, 1050: metapneustic, and 1051: apneustic (after Krivosheina)

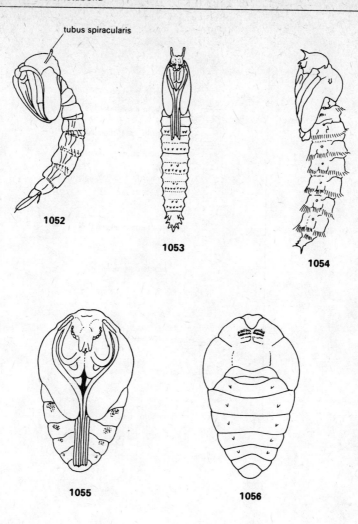

Figs. 1052–1056. Some types of Diptera pupa. – 1052: Aedes in lateral, 1053: Tipula in ventral, 1054: Bombylid in lateral, 1055–1056: Bibiocephala in ventral and dorsal view (after Peterson)

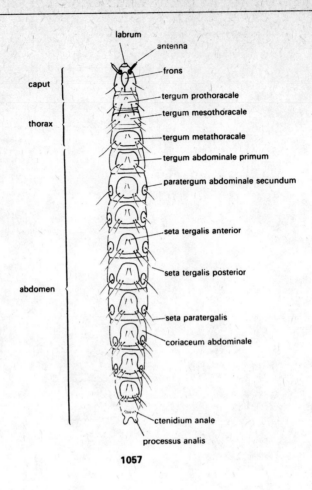

Fig. 1057. Schematic drawing of the Siphonaptera in dorsal view (after Bartkowska)

Holomorphosis II: Holometabolia

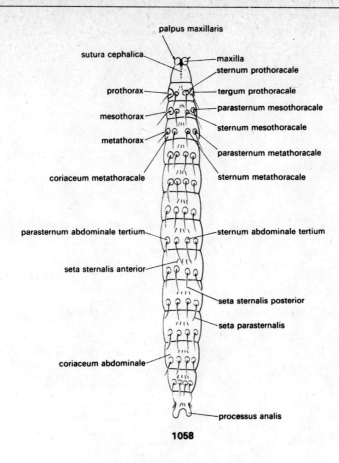

Fig. 1058. Schematic drawing of the Siphonaptera in ventral view (after Bartkowska)

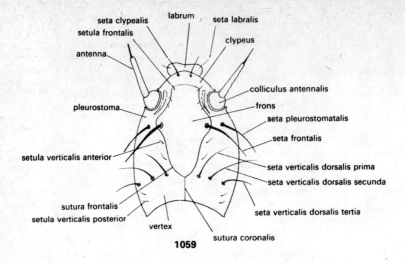
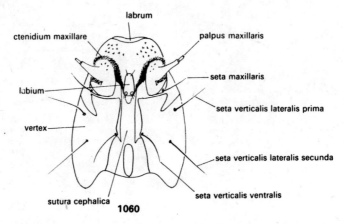

Figs. 1059–1060. Head of Siphonaptera in dorsal and ventral view (after Beier)

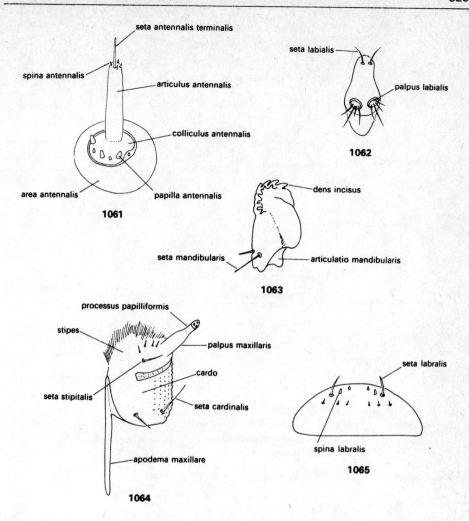

Figs. 1061—1065. Parts of head and mouthparts of Siphonaptera. – 1061: The antenna, 1062: the labium, 1063: the mandible, 1064: the maxilla, and 1065: the labrum (after Bartkowska)

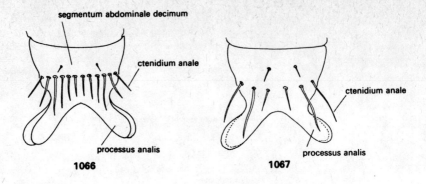
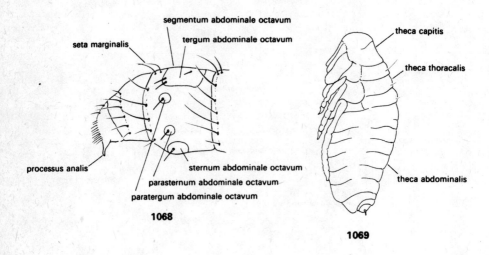

Figs. 1066–1068. Abdominal end of Siphonaptera. – 1066–1067: Typhloceras-type in dorsal and ventral, 1068: Tunga-type in lateral view. – Fig. 1069. The pupa (after Bartkowska)

INDEXES

LATIN–ENGLISH

The bold face numbers refer to figures in the two general parts.

abdomen – abdomen 1, **19, 20, 69,** 70, 90, 109, 126, 160, 195, 224, 250, 267, 285, 297, 313, 334, 346, 365, 383, 404, 418, 439, 473, **505–508, 527,** 529, 635, 636, 642, 655, 727, 743, 752, 788, 880, 1057
acanthoparia – acanthoparia 566
acroparia – acroparia 566
acrotergum – acrotergite 427
– mesothoracale – mesothoracic acrotergite **51,** 97, 100, 101, 357, 434, 780
– metathoracale – metathoracic acrotergite **54, 55,** 103, 104, 359
– prothoracale – prothoracic acrotergite **48, 49,** 97, 98, 258, 780
adfrons – adfrons **512, 513,** 790, 901, 902
admentum – admentum 92, 93
alveolus – alveole **41**
– anterior – anterior alveole 904
– frontalis – frontal alveole 901
– genalis – genal alveole 904
– labralis – labral alveole 917
– lateralis – lateral alveole 904
– ocellaris – ocellar alveole 904
– posterior primus – first posterior alveole 901, 904
– – secundus – second posterior alveole 901, 904
– subocellaris – subocellar alveole 904
ampulla abdominalis dorsalis – dorsal abdominal ampulla 627
– – ventralis – ventral abdominal ampulla 789
– segmenti secundi – ampulla of second segment 507
– – septimi – ampulla of seventh segment 507
– thoracalis dorsalis – dorsal thoracic ampulla 627
– ventralis – ventral ampulla 713
anapleuron – anapleurite **46, 525**
– mesothoracale – mesothoracic anapleurite 12, 13, 101, **520**
– metathoracale – metathoracic anapleurite 15, 104, **521**
– prothoracale – prothoracic anapleurite 8, 9, **519**
anepimeron – anepimeral sclerite 427
– mesothoracale – anepimeral sclerite of mesothorax 434, 435
– metathoracale – anepimeral sclerite of metathorax 434, 435
– prothoracale – anepimeral sclerite of prothorax 279, 435, 612
anepisternum – anepisternal sclerite **47,** 427

anepisternum prothoracale – anepisternal sclerite of prothorax **49,** 279, 776
annulus → anulus
anteclypeus – anteclypeus 313, 366, 367, 481–485, 487, **512,** 730, 744, 790, 792, 903, 1022
antenna – antenna **19–22, 40,** 70–73, 76, 90–92, 94, 109, 110, 126, 130, 133, 135, 160–166, 169, 175, 195–198, 214, 224, 225, 250, 251, 267, 268, 285, 288, 297, 300, 313, 334, 335, 346–350, 365, 367, 383–385, 387, 404, 405, 416, 417, 439, 473, 477, 478, 482, 485, 490, **505, 512, 513, 515,** 547, 551–553, 555–560, 597, 648, 655, 671–675, 677, 690, 727, 730, 743–745, 752, 757, 758, 760, 763–765, 782, 788, 790–792, 794, 901, 903, 904, 943, 946, 988, 999, 1020–1022, 1025–1029, 1031, 1034, 1057, 1059
antennarium – antennal sclerite 670, 672, 678, 679
antennifer – antennifer **32,** 271, 581
anulus abdominalis – abdominal ring **506,** 618, 783, 799, 1007
– – tertius – third abdominal ring 701
– mesothoracalis – mesothoracic ring 597
– – primus – first mesothoracic ring 599, 944
– – quartus – fourth mesothoracic ring 599
– – secundus – second mesothoracic ring 599, 944
– – tertius – third mesothoracic ring 599, 944
– metathoracalis – metathoracic ring 597
– – primus – first metathoracic ring 945
– – secundus – second metathoracic ring 843, 945
– prothoracalis – prothoracic ring 597
– – primus – first prothoracic ring 841, 943
– – secundus – second prothoracic ring 943
– thoracalis primus – first thoracic ring **525**
– – secundus – second thoracic ring 693
anus – anus 184, 504, 614, 618, 715, 717, 720, 983, 1042, 1043
apex antennalis – apex of antenna **40,** 386, 457, 679
– branchialis – apex of tracheal gill 148, 189
– caudalis – caudal apex 70
– galealis – apex of galea **35**
– glossalis – apex of glossa **39**
– lacinialis – apex of lacinia **36**
– lobi molaris inferior – apex of lower molar lobe **26**
– – – superior – apex of upper molar lobe **26**
– mandibularis – apex of mandible **26,** 134, 388, 824
– palpi labialis – apex of labial palp **37**
– – maxillaris – apex of maxillary palp **34,** 255
– paraglossalis – apex of paraglossa **38**
– unguis – apex of claw 85
apodema maxillare – maxillary apodeme 1064
apophysis tarsalis – tarsal apophysis 414
apotorma – apotorma 566
appendicula analis – anal appendage 620

appendicula dorsalis – dorsal appendage 184, 185
– sensillaris – sensory appendage 583
– subanalis – subanal appendage 717
– ventralis – ventral appendage 184, 185
appendix abdominalis – abdominal appendage 107, 108
area analis – anal area 1040
– antennalis – antennal area 796, 1061
– denticulata – denticulate area 1020
– dorsalis – dorsal area 879, 880, 1040
– infraspiracularis – infraspiracular area 701
– intermedialis – intermedial area 1040
– lateralis – lateral area 1040
– medialis – median area 1040
– molaris – molar area 75, 137
– ocellaris – ocellar area 70–73, 76, 419–421, **513**, 635
– ocularis – ocular area 790
– paraspiracularis – paraspiracular area 701
– postspiracularis – postspiracular area 701, 717
– sensillaris – sensory area 581, 916
– spiracularis – spiracular area 701, 717
– subdorsalis – subdorsal area 879, 880, 1040
– subspiracularis – subspiracular area 701, 702, 704, 711, 717, 879
– suprapedis – suprapedal area **528**, 701, 702, 717
– supraspiracularis – supraspiracular area 701, 879, 880
– ventralis – ventral area 879, 1040
areola antennalis – antennal areola 426
– pedicellaris – pedicellar areola 423
arolium – arolium **66**, 263, 429–433, 698, 699, 772–775
articulatio branchialis – articulation of tracheal gill 148
– coxalis – coxal articulation **58, 59**
– mandibularis – articulation of mandible 1030, 1063
articulus antennalis – antennal joint **41**, 488, 679, 916, 1061
– – anularis – annular joint of antenna **40**
– – apicalis – apical joint of antenna **40**, 489, 579, 678, 746
– – cylindricus – cylindrical joint of antenna **40**
– – primus – first antennal joint 546, 583, 587
– cercalis – cercal joint 221
– – apicalis – apical joint of cercus 364
– – basalis – basal joint of cercus 364
– rostralis apicalis – apical joint of rostrum 454
– – primus – first joint of rostrum 454
– – secundus – second joint of rostrum 454
– – tertius – third joint of rostrum 454
artus apicalis – apical joint 797
– basalis – basal joint 797
– branchialis apicalis – apical joint of tracheal gill 740

artus branchialis basalis – basal joint of tracheal gill 740
– – medialis – middle joint of tracheal gill 740
– medialis – middle joint 797
– palpi labialis apicalis – apical joint of labial palp **37,** 771, 932
– – – basalis – basal joint of labial palp **37,** 771, 932
– – – medialis – middle joint labial palp **37,** 771
– – maxillaris apicalis – apical joint of maxillary palp **34,** 77, 931
– – – basalis – basal joint of maxillary palp **34,** 768, 931
– – – medialis – middle joint of maxillary palp **34,** 235
auxilium – auxilium **66,** 394

barbula – barbula 614, 616
basicardo – basicardo 561, 563, 565
basicoxa – basicoxite **47,** 59–61
– mesothoracalis – mesothoracic basicoxite 259
– metathoracalis – metathoracic basicoxite 259
basimentum – basimentum 806
basisternum – basisternum **46, 47,** 58
– mesothoracale – mesothoracic basisternum 13, **53,** 81, 102, 117, 182, 209, 237, 240, 259, 294, 311, 344, 358, 399, 462–464, 491, **523,** 691, 781, 845, 944
– metathoracale – metathoracic basisternum 16, **56,** 83, 105, 118, 182, 210, 237, 240, 259, 295, 345, 360, 400, 464, **524,** 692, 781, 846, 945
– prothoracale – prothoracic basisternum 10, **50,** 79, 98, 99, 116, 208, 237, 240, 257, 293, 307, 343, 356, 398, **522,** 690, 781, 844, 943
basistipes – basistipes 271–273, 317, 389, 390
basivalvula – basivalvula 249
branchia – tracheal gill 742
– abdominalis – abdominal tracheal gill 150, 196, 540, 727, 799
– – dorsalis – dorsal abdominal tracheal gill 147, 870
– – lateralis – lateral abdominal tracheal gill 149, 804
– – septima – seventh abdominal tracheal gill 738
– – ventralis – ventral abdominal tracheal gill 804, 870
– analis – anal tracheal gill 628, 799, 853, 854, 999, 1011, 1042, 1044
– caudalis – caudal tracheal gill 161
– cercalis – cercal tracheal gill 161
– coxalis mesothoracalis – mesothoracic coxal tracheal gill 211
– – metathoracalis – metathoracic coxal tracheal gill 211, 214
– – prothoracalis – prothoracic coxal tracheal gill 208, 214
– mesopleuralis – mesopleural tracheal gill 209
– metapleuralis – metapleural tracheal gill 210
– metathoracalis – metathoracic tracheal gill 727
– pleuralis – pleural tracheal gill 212
– prothoracalis – prothoracic tracheal gill 145, 146
– rectalis – rectal tracheal gill 994
– sternalis – sternal tracheal gill 213

branchia subcoxalis – subcoxal tracheal gill 144
– subspiracularis – subspiracular tracheal gill 971
– supraspiracularis – supraspiracular tracheal gill 971
buccula – buccula 454, 459, 482, 485, 486

calcar spiraculare – spiracular spur 629
– tibiale – tibial spur **57, 65, 66**, 308, 309, 376
campus – campus 614–617
canalis alimentarius – alimentary canal 762
– mandibularis – mandibular canal 576
– salivarius – salivary canal 933, 934
– venenatus – poison canal 762
caput – head 1, **19–24**, 70, 90, 109, 126, 160, 195, 214, 224, 250, 267, 285, 297, 313, 334, 346, 365, 383, 404, 418, 439, 473, **505–508, 512–515,** 529, 635–637, 642, 655, 727, 743, 752, 788, 799, 879, 880, 988, 999, 1001, 1027–1029, 1057
cardo – cardo 4, 6, **22, 24, 34,** 77, 95, 111, 113, 136, 177, 198, 226, 227, 233–235, 252, 253, 255, 271–273, 287, 299, 301, 317–320, 336, 350, 352, 369, **514, 515,** 547, 560, 561, 563–565, 597, 677, 728, 729, 734, 745, 759, 761, 763, 765, 768, 813, 1020, 1023, 1024, 1027, 1064
carina coxalis – coxal carina 308
– femoralis – femoral carina 309
– – dorsalis – dorsal femoral carina **64**
– – ventralis – ventral femoral carina **64**
– occipitalis – occipital carina 197
– unguis – ungual carina 263
cauda – tail 90, 415, 437, 438, 504, **505, 506,** 784, 884
cavea antennalis – antennal socket **21, 32,** 251, 298, 347, 353, 366, 481, 791, 799, 805
– cardinalis – cardinal cavity **34**
– centralis – central cavity **528**
– coxalis anterioris – fore coxal cavity **43**, 293, 307, **522,** 612, 781, 954
– – apicalis – apical coxal cavity **60, 61**
– – basalis – basal coxal cavity **60, 61**
– – medialis – middle coxal cavity **44**, 294, **523,** 612, 781, 955
– – posterioris – hind coxal cavity **45**, 295, **524,** 612, 781, 956
– femoralis apicalis – apical femoral cavity **64**
– – basalis – basal femoral cavity **64**
– galealis – galeal cavity **35**
– glossalis – glossal cavity **39**
– lacinialis – lacinial cavity **36**
– mandibularis – mandibular cavity **26**
– paraglossalis – paraglossal cavity **38**
– postmentalis – postmental cavity **37**
– praementalis – premental cavity **37**

cavea stipitalis – stipital cavity **34**
– trochanteralis apicalis – apical trochanteral cavity **62, 63**
– – basalis – basal trochanteral cavity **62, 63**
caverna basicoxalis – basicoxal cavity **60**
– sensillaris – sensory pit **768**
cephalotheca – cephalotheca 641, 647
cephalothorax – cephalothorax 479
cerculus – cerculus 184, 185
cercus – cercus **19, 69**, 90, 108, 109, 125, 126, 147, 149, 195, 196, 217–219, 224, 248–250, 266–268, 282, 285, 296, 313, 334, 346
cervix – cervix 270, 324, 326, **522**
chaetoparia – chaetoparia 566
chalaza – chalaza 906
cibarium – cibarium 1025, 1027
cilium – thin hair 906
– tibiale – tibial hair **65**
clavus branchialis externus – external margin of tracheal gill 189
– – internus – internal margin of tracheal gill 189
– clypealis apicalis – apical margin of clypeus **29**
– – basalis – basal margin of clypeus **29**
– – lateralis – lateral margin of clypeus **29**
– clypeolabralis apicalis – apical margin of clypeolabrum **30**
– – basalis – basal margin of clypeolabrum **30**
– – lateralis – lateral margin of clypeolabrum **30**
– coxalis apicalis – apical margin of coxa **60, 61**
– – basalis – basal margin of coxa **60, 61**
– femoralis dorsalis – dorsal margin of femur **64**
– – ventralis – ventral margin of femur **64**
– frontalis apicalis – apical margin of frons **27**
– – basalis – basal margin of frons **27**
– – lateralis – lateral margin of frons **27**
– frontoclypealis apicalis – apical margin of frontoclypeus **28**
– – basalis – basal margin of frontoclypeus **28**
– – lateralis – lateral margin of frontoclypeus **28**
– labralis apicalis – apical margin of labrum **31**
– – basalis – basal margin of labrum **31**
– – lateralis – lateral margin of labrum **31**
– lobi lateralis exterior – outer margin of lateral lobe 174
– – – inferior – lower margin of lateral lobe 174
– – – superior – upper margin of lateral lobe 174
– mesonotalis anterior – fore margin of mesonotum **44, 517**
– – lateralis – lateral margin of mesonotum **44, 517**
– – posterior – hind margin of mesonotum **44, 517**
– mesopleuralis anterior – fore margin of mesopleuron **44**
– – dorsolateralis – dorsolateral margin of mesopleuron **44**
– – posterior – hind margin of mesopleuron **44**

Latin—English

clavus mesopleuralis ventrolateralis – ventrolateral margin of mesopleuron **44**
– mesosternalis anterior – fore margin of mesosternum **44**
– – lateralis – lateral margin of mesosternum **44**
– – posterior – hind margin of mesosternum **44**
– metanotalis anterior – fore margin of metanotum **45, 518**
– – lateralis – lateral margin of metanotum **45, 518**
– – posterior – hind margin of metanotum **45, 518**
– metapleuralis anterior – fore margin of metapleuron **45**
– – dorsolateralis – dorsolateral margin of metapleuron **45**
– – posterior – hind margin of metapleuron **45**
– – ventrolateralis – ventrolateral margin of metapleuron **45**
– metasternalis anterior – fore margin of metasternum **45**
– – lateralis – lateral margin of metasternum **45**
– – posterior – hind margin of metasternum **45**
– ocularis anterior – anterior ocular margin **21**
– – posterior – posterior ocular margin **21**
– pleurostomatalis anterior – fore margin of pleurostoma **33**
– – apicalis – apical margin of pleurostoma **33**
– – basalis – basal margin of pleurostoma **33**
– – posterior – hind margin of pleurostoma **33**
– postoccipitalis orbitalis — orbital margin of postocciput **25**
– pronotalis anterior – fore margin of pronotum **43, 516**
– – apicalis – apical margin of pronotum 827
– – basalis – basal margin of pronotum 826
– – dorsolateralis – dorsolateral margin of pronotum **43**
– – lateralis – lateral margin of pronotum **43, 516**
– – posterior – hing margin of pronotum **43, 516**
– propleuralis anterior – fore margin of propleuron **43**
– – dorsolateralis — dorsolateral margin of propleuron **43**
– – posterior – hind margin of propleuron **43**
– – ventrolateralis – ventrolateral margin of propleuron **43**
– prosternalis anterior – fore margin of prosternum **43**
– – lateralis – lateral margin of prosternum **43**
– – posterior – hind margin of prosternum **43**
– trochanteralis apicalis – apical margin of trochanter **62, 63**
– – basalis – basal margin of trochanter **62, 63**
clithrum – clithrum 566
clypeolabrum – clypeolabrum **30**, 272, 405, 419, 763, 791
clypeus – clypeus 2, **21, 22, 29,** 71–73, 110, 131, 135, 162, 197, 230, 231, 251, 270, 286, 288, 298, 300, 335, 347–349, 421, 454, 486, **513,** 560, 595, 648, 672, 673, 675, 690, 794, 901, 1019, 1021, 1059
– dorsalis – dorsal clypeus 384
– ventralis – ventral clypeus 385
colliculus antennalis – antennal mount 1059, 1061
collum – neck 8, 435, 672, 781
condylus mandibularis — mandibular condyle 674, 731, 795, 824, 921

condylus mandibularis anterior – anterior condyle of mandible **33**
– – apicalis – apical condyle of mandible 388
– – basalis – basal condyle of mandible 388
– – posterior – posterior condyle of mandible **33**
conus – cone 405, 406
coriaceum abdominale – abdominal skin 1057, 1058
– mesothoracale – mesothoracic skin **517**
– metathoracale – metathoracic skin **518**, 1058
– prothoracale – prothoracic skin **516**
– thoracale – thoracic skin 799
corium abdominale intersegmentale – intersegmental membrane of abdomen 529
– antennale – antennal membrane **41**, 225, 269, 271, 546, 587, 678, 902, 903, 1025
– cervicale – cervical membrane **25**
– clypeale – clypeal membrane 902
– coxale – coxal membrane **58, 59, 525**
– femorale – femoral membrane 949
– mandibulare – mandibular membrane 253, 560, 672, 676, 1019
– mentale – mental membrane 729
– palpi labiale – membrane of labial palp 236
– – maxillare – membrane of maxillary palp 233
– pedicellare – pedicellar membrane **41**
– pleurale – pleural membrane **525**
– prothoracale – prothoracic membrane 546
– scapale – scapal membrane **41**
– stipitale – stipital membrane 561
– tarsale – tarsal membrane **526**, 950
– tibiale – tibial membrane **526**, 950
– unguis – ungual membrane 377
cornu anale – anal horn 984
corona molaris – molar crown **26**, 921
corpotentorium – corpotentorium 676
corpus – corpus 89
corypha – corypha 566
– verticalis – vertical tuft 805
coxa – coxa 17, **57–59**, 84, 106, 119–121, 144, 215, 241, 243, 245, 281, 308, 331, 332, 338, 374, 375, 394, 396, 427, 428, 466, 467, 498, **525**, 601–605, 693, 694, 697, 735, 748, 847–850, 949, 950
– anterior – fore coxa 3, 8–10, **42, 49, 50,** 78, 79, 97–99, 116, 145, 146, 163, 164, 175, 182, 183, 208, 232, 237, 239, 240, 257, 258, 260, 279, 280, 290, 293, 306, 310, 321, 324–326, 340, 343, 354, 356, 378, 381, 398, 401, 407, 435, 436, 462, 483, 484, **519, 522,** 598, 608, 609, 690, 776, 841, 844
– fixa – fixed coxa **60**
– libera – free coxa **61**
– medialis – middle coxa 12, 13, **42, 52, 53,** 80, 81, 100–102, 117, 164, 183, 209, 237–240, 247, 259, 261, 264, 279, 280, 291, 294, 310, 311, 327, 328, 341, 344,

358, 379, 381, 399, 402, 407, 435, 436, 463–465, 491, **520, 523,** 599, 608, 609, 777, 842, 845
– posterior – hind coxa 15, 16, **42, 55, 56,** 82, 83, 105, 118, 183, 210, 216, 237–240, 247, 259, 262, 264, 279, 280, 292, 295, 311, 329, 330, 342, 345, 360, 380, 381, 400, 403, 407, 435, 436, 464, 465, 492, **521, 524,** 600, 608, 609, 778, 843, 846
coxopleuron – coxopleurite **525,** 778
– mesothoracale – mesothoracic coxopleurite **520, 523,** 842, 845, 944, 947
– metathoracale – metathoracic coxopleurite **521, 524,** 843, 846, 945, 948
– prothoracale – prothoracic coxopleurite **519,** 841, 943, 946
coxopodium – coxopodite 108, 738, 739, 742
coxosternum abdominale – abdominal coxosternite 123, 248, 266
crepis – crepis 566
crista frontalis – frontal crest 550
– frontoverticalis – frontovertical crest 546
– pronotalis – pronotal crest **48**
– – longitudinalis – longitudinal crest of pronotum 224
– verticalis – vertical crest 546
ctenidium anale – anal comb 983, 1057, 1066, 1067
– branchiale – comb of tracheal gill 148, 189
– caudale – caudal comb 858
– galeale – galeal comb **35**
– glossale – glossal comb **39**
– labiale – labial comb 1026
– laciniale – lacinial comb **36,** 204, 318, 352
– maxillare – maxillary comb 1060
– paraglossale – paraglossal comb **38**

dens – dens 88, 89
– dorsalis – dorsal tooth 920
– incisus – incisor **26,** 74, 112, 176, 370, 684, 731, 747, 823, 1030, 1063
– subapicalis – subapical tooth **26,** 112
– ventralis – ventral tooth 920
dentes mediales – incisor lobe 920
denticulus apicalis – apical denticle 88, 170, 171, 173
– basalis – basal denticle 88
– dorsalis – dorsal denticle 85
– femoralis – femoral denticle 498
– galeolacinialis – denticle of galeolacinia 562
– laciniae apicalis – apical denticle of lacinia 235, 318
– – externus – external denticle of lacinia 204
– – internus – internal denticle of lacinia 204
– – medialis – middle denticle of lacinia 235
– – subapicalis – subapical denticle of lacinia 234, 235, 318
– lateralis – lateral denticle 85
– mandibularis – mandibular denticle 388, 767

denticulus mobilis – movable hook 165, 170–174
- pollicis – denticle of pollex 413
- pronotalis – pronotal denticle 305
- subapicalis – subapical denticle 88, 170, 172, 173
- tibialis – tibial denticle **65**
- unci – denticle of uncus 854
- unguis – denticle of claw 377
- – basalis – basal denticle of claw 606
- – subapicalis – subapical denticle of claw 606
- ventralis – ventral denticle 85
dexiotorma – dexiotorma 566
discus analis – anal disc 504
- caudalis – caudal disc 998
dististipes – dististipes 273
dorsocoxa – dorsocoxa **60, 61**, 526, 951

empodium – empodium 85, 122, 696
epicranium – epicranium 72, 91–94, 251–253, 313, 320, 348–350, 353, 405, **512–515**, 547, 555, 557, 595, 596, 792, 793, 903, 904, 1023–1026
epifrons – epifrons **21, 22, 32**
epimeron – epimeron **47, 58, 59**, 332, **525**
- mesothoracale – mesothoracic epimeron **52,** 247, 294, 327, 328, 341, 379, 464, **520,** 612, 842
- metathoracale – metathoracic epimeron **55,** 247, 295, 329, 330, 342, 380, 464, **521,** 612, 843
- prothoracale – prothoracic epimeron **49,** 258, 293, 307, 321, 325, 340, 354, 378, 462, **519,** 776, 841
epipharynx – epipharynx 648
epipleuron – epipleuron 954
- abdominale octavum – eighth abdominal epipleuron 618
- – primum – first abdominal epipleuron 596
- – septimum – seventh abdominal epipleuron 618
- mesothoracale – mesothoracic epipleuron 596
- metathoracale – metathoracic epipleuron 596, 843
epiproctum – epiproct 161, 382
episternum – episternum **58, 59,** 332, **525**
- abdominale – abdominal episternum 613
- mesothoracale – mesothoracic episternum **52,** 294, 327, 328, 341, 379, 434, 435, 464, **520,** 612
- metathoracale – metathoracic episternum **55,** 247, 295, 329, 330, 342, 380, 434, 435, 464, **521,** 612, 842, 843
- prothoracale – prothoracic episternum 258, 293, 307, 321, 325, 340, 343, 354, 378, 435, 462, **519,** 612, 841
epistoma – epistoma 674, 675
epitorma – epitorma 566
epizygum – epizygum 566

euplantula – euplantula **66**, 246, 394
eupleuron – dorsal supracoxal sclerite **46**
– prothoracale – dorsal supracoxal sclerite of prothorax 98
eusternum – eusternum **46**
eustipes – eustipes 317, 561

femur – femur 17, **57, 64,** 84, 106, 119, 120, 144, 215, 241–245, 281, 308, 309, 331, 338, 363, 374, 375, 394–396, 428, 466–469, 471, 498, 501, **525, 526,** 601–605, 693, 694, 697, 735, 748, 772, 847–850, 949–951
– anterius – fore femur **42,** 141, 205, 290, 392, 461, 462, 598, 776
– mediale – middle femur **42,** 142, 205, 259, 291, 392, 463, 599, 777
– posterius – hind femur **42,** 143, 205, 259, 292, 324, 392, 600, 749, 778
filamentum abdominale – abdominal filament 742, 882
– – octavum – eighth abdominal filament 738
– – primum – first abdominal filament 736, 737
– – septimum – seventh abdominal filament 738, 741
– capitis – head filament 885
– caudale – caudal filament 727
– intestinale – intestinal filament 853
– pygopodii – filament of pygopod 738, 739, 741
filum terminale – terminal filament 109, 125, 126, 147, 150, 858
fissura labii – labial fissure 5
– mentalis – mental fissure 371
– praementalis – premental fissure 201, 302, 314
– submentalis – submental fissure 813
flagellomeron – flaggellar joint 275
flagellum – flagellum **40,** 386, 769
floccus – tuft 998
– genalis – genal tuft 131
foramen occipitale – occipital foramen **23–25,** 76, 226, 252, 287, 299, 303, 320, 336, 368, **513–515,** 552, 556, 557, 676, 728, 729, 759, 793, 806, 1026
– prothoracale – prothoracic foramen 307
– thoracale – thoracic foramen 609, 781
forceps – forceps 267
fossa antennae – antennal furrow 387
– basicoxalis – basicoxal furrow **60**
– clypealis transversalis – transversal furrow of clypeus 670
– clypeolabralis longitudinalis – longitudinal furrow of clypeolabrum **30**
– – transversalis – transversal furrow of clypeolabrum **30**
– coxalis – coxal furrow **60**
– femoralis – femoral furrow **64,** 215, 309
– frontalis longitudinalis – longitudinal furrow of frons **27**
– – transversalis – transversal furrow of frons **27**
– frontoclypealis longitudinalis – longitudinal furrow of frontoclypeus **28**
– – transversalis – transversal furrow of frontoclypeus **28**
– galealis – galeal furrow **35**

fossa glossalis – glossal furrow **39**
– labralis – labral furrow 818
– lacinialis – lacinial furrow **36**
– mandibularis – mandibular furrow **26**, 568, 767, 823
– mesosternalis sagittalis – sagittal furrow of mesosternum 13
– paraglossalis – paraglossal furrow **38**
– pronotalis – pronotal furrow 826
– rostralis – rostral furrow 459
– tibialis – tibial furrow **65**
– verticalis – vertical furrow **21**, 670, 673, 676, 790
fovea tentorialis – tentorial pit 286
– – anterior – anterior tentorial pit 91, 648, 675, 730, 792
– – posterior – posterior tentorial pit 92, 648, 729, 793
– – superior – superior tentorial pit 91, 648, 672
frons – frons 2, **21, 22, 27,** 72, 73, 110, 135, 162, 166, 167, 169, 197, 225, 230–232, 251, 269, 270, 272, 286, 288, 298, 300, 313, 335, 347, 366, 367, 384, 419–421, 458, 485, **512, 513,** 546, 548, 549, 551, 560, 595, 648, 670, 673, 675, 690, 730, 744, 790–792, 794, 901–904, 1019, 1021–1024, 1057, 1059
frontoclypeus – frontoclypeus **28,** 91, 94, 133, 271, 460, 760, 805
furca – furca 81
furcasternum – furcasternum **46, 47**
– mesothoracale – mesothoracic furcasternum 12, **53,** 81, 102, 117, 182, 209, 237, 240, 259, 294, 311, 358, 463–465, 491, 691, 781
– metathoracale – metathoracic furcasternum **56,** 83, 105, 118, 182, 210, 237, 240, 259, 280, 295, 360, 464, 692
– – laterale – lateral furcasternum of metathorax 16
– – mediale – middle furcasternum of metathorax 16
– prothoracale – prothoracic furcasternum 10, **50,** 79, 98, 99, 116, 208, 237, 240, 257, 293, 307, 326, 690, 781
fusulus – spinneret 813, 903, 918, 933
– denticulatus – denticulate spinneret 934

galea – galea 4, 6, **22, 24, 34, 35,** 95, 111, 113, 177, 198, 202, 204, 226, 231, 233–235, 252, 253, 255, 270–273, 287–289, 299, 301, 317–320, 336, 352, 368, 369, 389, 390, **513,** 547, 561, 563, 670–672, 728, 729, 732, 734, 791, 930, 931, 1020
galeolacinia – galeolacinia 227, 229, 562, 793
gena – gena **21–23,** 73, 110, 132, 163, 164, 175, 229–231, 271, 286, 288, 298, 300, 335, 366, 3667, 383, 458, 481–483, 485–487, 560, 648, 670, 674, 677, 728, 765, 791, 1023
– dorsalis – dorsal gena 384, 387
– ventralis – ventral gena 385
genu – genu **57, 65,** 241, 242, 331, 493, 748, 847
glandula abdominalis dorsalis – dorsal abdominal gland 630, 968
– salivaria – salivary gland 762
– venenata – poison gland 762, 766

glossa – glossa 5, **22, 24, 37, 39,** 93, 114, 173, 199–201, 226–228, 236, 252, 256, 287, 289, 302, 314, 320, 336, 351, 368, 373, 391, **513**
gula – gula **23, 24,** 71, 72, 198, 226, 227, 252, 256, 271, 273, 287, 350, 368, 454, 458, 459, **513–515,** 554–557, 671, 676, 728, 729, 745, 806, 813–815, 817, 1025, 1026
gulamentum – gulamentum 314, 320, 385
gymnoparia – gymnoparia 566

halter – halter 649, 654
hamulus – hamule **528,** 786
– propodii – crochet of abdominal leg 982
hamus – hook 406
– caudalis – caudal hook 619
– dorsalis – dorsal hook 626
– pseudopodii – crochet of pseudopod 988
– pygopodii – hook of pygopod 739, 741
haptolachus – haptolachus 566
haptomerum – haptomerum 566
helus – helus 566
hemisternum abdominale – abdominal hemisternite 364
hemitergum abdominale – abdominal hemitergite 346
hypopharynx – hypopharynx **22,** 72, 73, 138, 198, 231, 546, 578, 648, 671, 672, 1024, 1025
hypopleuron – hypopleuron 954
– abdominale octavum – eighth abdominal hypopleuron 618
– – primum – first abdominal hypopleuron 596
– – septimum – seventh abdominal hypopleuron 618
– mesothoracale – mesothoracic hypopleuron 596
– metathoracale – metathoracic hypopleuron 596, 843
– prothoracale – prothoracic hypopleuron 596
hypopodium – hypopodite 528, 701, 972
hyposternum abdominale – abdominal hyposternum 613
hypostoma – hypostoma 674, 675, 729

interpleuron mesothoracale – mesothoracic interpleurite 379, 381
intersternum – intersternite 92
iugulum – jugulum 237
iugum – jugum 454, 458–460, 482, 485–487

j → i

katapleuron – katapleurite **46**
– mesothoracale – mesothoracic katapleurite 12, 101
– metathoracale – metathoracic katapleurite 15, 104
– prothoracale – prothoracic katapleurite 9
katepimeron – lower epimeral sclerite 427
– mesothoracale – lower epimeral sclerite of mesothorax 434, 435

katepimeron metathoracale – lower epimeral sclerite of metathorax 434, 435
– prothoracale – lower epimeral sclerite of prothorax 240, 279, 612
katepisternum – lower episternal sclerite **47**, 427
– prothoracale – lower episternal sclerite of prothorax **49**, 279, 776

labium – labium **22, 37,** 71–73, 76, 110, 161, 229–231, 268, 271, 288, 353, 367, 406, 419, 421, 459, 481–486, **513,** 546, 578, 690, 759, 765, 793, 903, 943, 946, 1020, 1026, 1060
labrum – labrum 2, **21, 22, 24, 31,** 71–73, 91, 92, 94, 110, 130–132, 135, 162, 175, 196–198, 225–227, 229–231, 251, 269–271, 273, 286, 288, 298, 300, 313, 335, 347, 349, 350, 353, 366–368, 385, 421, 454, 459, 483–487, **512–514,** 560, 567, 595, 597, 648, 655, 670–675, 690, 728, 730, 743, 744, 759, 790, 792, 794, 799, 805, 901–904, 943, 946, 1019, 1020, 1022–1027, 1057, 1059, 1060
lacinia – lacinia 4, 6, **22, 24, 34, 36,** 95, 111, 113, 177, 198, 202, 204, 226, 231, 233–235, 255, 271, 273, 289, 299, 301, 317, 319, 320, 336, 352, 368, 369, 389, 390, **513,** 561, 563, 672, 728, 734, 745, 768, 930, 931, 1020
laeotorma – laeotorma 566
lamina analis – anal plate 972
– antennalis – antennal appendage 40
– branchialis – plate of tracheal gill 148
– – apicalis – apical plate of tracheal gill 189
– – basalis – basal plate of tRracheal gill 189
– cervicalis – cervical sclerite 9, 398
– – apicalis – apical cervical sclerite 293
– – basalis – basal plate of tracheal gill 189
– – dorsalis – dorsal cervical sclerite **49,** 183, 337, 340, 347, 354, 355
– – lateralis – lateral cervical sclerite 8, **49,** 257, 321, 340, 343, 350, 354, 356, 378, 677
– – medialis – median cervical sclerite 8, 183, 677
– – ventralis – ventral cervical sclerite **49,** 175, 183, 257, 321, 340, 350, 354, 356
– dorsalis – dorsal plate 503
– galealis – galeal plate **35**
– glossalis – glossal plate **39**
– infraanalis – infra-anal plate 184, 185
– intercervicalis – intercervical sclerite 303
– labii – labial plate 3
– lacinialis – lacinial plate **36**
– laterocervicalis anterior – anterior laterocervical sclerite 303
– – medialis – median laterocervical sclerite 303
– – posterior – posterior laterocervical sclerite 303
– mediocervicalis anterior – anterior mediocervical sclerite 303
– – posterior – posterior mediocervical sclerite 303
– mesonotalis lateralis anterior – anterior lateral plate of mesonotum 840
– – – posterior – posterior lateral plate of mesonotum 840
– – medialis – medial plate of mesonotum 840
– – mediolateralis anterior – anterior mediolateral plate of mesonotum 840
– – – posterior – posterior mediolateral plate of mesonotum 840

lamina metanotalis lateralis – lateral plate of metanotum 840
– – medialis – medial plate of metanotum 840
– – mediolateralis anterior – anterior mediolateral plate of metanotum 840
– – – posterior – posterior mediolateral plate of metanotum 840
– paraglossalis – paraglossal plate **38**
– postantennalis – postantennal sclerite 385
– postspiracularis – postspiracular sclerite 341
– – mesothoracalis – postspiracular sclerite of mesothorax **52**
– – metathoracalis – postspiracular sclerite of metathorax **55**
– pronotalis – pronotal plate 305
– – lateralis – lateral pronotal plate 840
– – medialis – median pronotal plate 840
– spiracularis – spiracular sclerite 608, 609
– subanalis – subanal plate 219, 266, 296, 382, 614, 617, 715, 717
– subantennalis – subantennal sclerite 385
– subspiracularis – subspiracular sclerite 258, 609, 709
– – mesothoracalis – subspiracular sclerite of mesothorax **52**
– – metathoracalis – subspiracular sclerite of metathorax **55**
– supraanalis – supra-anal plate 184, 217–219, 621, 715, 717
– suprapedis – suprapedal sclerite 709
– ventralis apicalis – apical ventral plate 503
– – basalis – basal ventral plate 503
– unguitractoralis – unguitractor plate 122, 432
laterosternum abdominale quartum – fourth abdominal laterosternite **527**
– mesothoracale – mesothoracic laterosternite 294
– metathoracale – metathoracic laterosternite 295
– prothoracale – prothoracic laterosternite 293
laterotergum abdominale – abdominal laterotergite 296
– – secundum – second abdominal laterotergite **527**
– mesothoracale – mesothoracic laterotergite 11
laterovertex – laterovertex 1022
ligula – ligula 92, 131, 132, 139, 140, 170, 299, 555, 557, 676, 728, 729, 745, 933
linea abdominalis – abdominal line **69, 506**
– addorsalis – addorsal line 878, 880
– clypealis – clypeal line **29**
– clypeolabralis longitudinalis – longitudinal line of clypeolabrum **30**
– – transversalis – transversal line of clypeolabrum **30**
– dorsalis – dorsal line 880
– dorsolateralis – dorsolateral line 750
– dorsopleuralis – dorsopleural line 527
– frontalis longitudinalis – longitudinal line of frons **27**
– – transversalis – transversal line of frons **27**
– frontoclypealis longitudinalis – longitudinal line of frontoclypeus **28**
– – transversalis – transversal line of frontoclypeus **28**
– genalis – genal line **32**

linea gularis longitudinalis – longitudinal line of gula 25
– – transversalis – transversal line of gula 25
– labralis – labral line 31
– lateralis – lateral line 800, 870, 879
– mesonotalis – mesonotal line **51**, 278, 828
– metanotalis – metanotal line 829
– pleuroventralis – pleuroventral line **527**
– pronotalis – pronotal line 205, 278, 827
– – anterior – anterior line of pronotum 305
– – longitudinalis – longitudinal line of pronotum 355
– – medialis – median line of pronotum 393
– – posterior – posterior line of pronotum 305
– – transversalis – transversal line of pronotum 355
– subdorsalis – subdorsal line 879, 880
– subventralis – subventral line 879
– ventralis – ventral line 76, 79, 81, 83, 87
lingua – lingua 92, 94, 115, 918
lobus analis – anal lobe 623, 1040
– antennalis – antennal lobe **40**
– coxalis – coxal lobe – **60, 61**
– externus maxillae – galea **514**
– femoralis dorsalis – dorsal lobe of femur **64**
– – ventralis – ventral lobe of femur **64**
– galealis – galeal lobe **35**
– genicularis – genicular lobe 242
– genitalis – genital lobe 284
– glossae externus – external lobe of glossa 115
– – internus – internal lobe of glossa 115
– glossalis – glossal lobe **39**
– internus – lacinia 806, 813
– – maxillae – lacinia **514**
– lacinialis – lacinial lobe **36**
– lateralis – lateral lobe 162–164, 166–168, 170, 173–175
– maxillaris – maxillary lobe 419, 420
– medialis – medial lobe 981
– molaris – molar lobe 254, 687
– paraglossae externus – external lobe of paraglossa 115, 299
– – internus – internal lobe of paraglossa 115, 299
– paraglossalis – paraglossal lobe **38**
– pronotalis anterior – anterior lobe of pronotum 169
– – posterior – posterior lobe of pronotum 169
– pygopodii – lobe of pygopod 853
– subanalis – subanal lobe 248, 266, 282
– supraanalis – supra-anal lobe 248
– trochanteralis anterior – anterior lobe of trochanter **62, 63**

lobus trochanteralis posterior – posterior lobe of trochanter **62, 63**
– unguis – ungual lobe 377
lunula femoralis – femoral lunule 242

macrochaeta caudalis – caudal bristle 437
mala – lobe 136
– labialis – labial lobe 1023
– maxillaris – maxillary lobe 555, 565, 1024
mandibula – mandible 5, **22, 24, 26,** 71–73, 92, 94, 110, 131–135, 175, 196–198, 225–227, 229, 230–232, 251–253, 269–273, 286, 288, 298, 299, 313, 320, 335, 336, 347, 349, 350, 353, 366–368, 385, 389, 406, **505, 512, 513, 515,** 546, 547, 556, 557, 559, 560, 595, 597, 643, 648, 670, 672, 674, 675, 677, 690, 728, 730, 744, 745, 752, 757–764, 790–794, 805, 806, 818, 901, 903, 943, 946, 988, 1019, 1022–1025, 1027–1029, 1031, 1034
– exodentate – exodentate mandible 1026
mandibulae – mandibles 782
manubrium – manubrium 89
maxilla – maxilla **34,** 71–73, 92, 94, 110, 132, 175, 229, 268, 288, **513,** 562, 648, 675, 690, 758–760, 762–764, 792, 918, 943, 946, 1019, 1026, 1058
membrana – membrane **67, 68**
– abdominalis – abdominal membrane 382
– – quarta – abdominal membrane of fourth segment 18
– intersegmentalis – intersegmental membrane 1, 437
– labralis – labral membrane 805
– mandibularis – mandibular membrane 388
– mentalis – mental membrane 701
membranaceum abdominale – abdominal skin **506,** 799
mentum – mentum **23,** 115, 200, 226–228, 236, 252, 253, 256, 270, 273, 299, 302, 371–373, 560, 561, 597, 648, 671, 677, 728, 729, 745, 806, 813, 903, 1023, 1024
meriston – meriston 276
meron – meron **47, 59**
mesofurca – mesofurca 344
mesonotum – mesonotum 1, 11, 12, **42, 44, 51, 52,** 70, 80, 90, 100, 101, 109, 117, 161, 178, 183, 211, 250, 260, 261, 267, 278, 280, 285, 294, 322, 327, 337, 341, 346, 357, 381, 383, 392, 393, 399, 404, 411, 416, 434, 435, 463, **517, 520,** 529, 596, 608–610, 612, 727, 736, 737, 743, 749, 752, 780, 782, 799, 842
mesopleuron – mesopleuron 12, **42, 44,** 80, 81, 183, 259–261, 280, 327, 381, 463, 491, 609
mesosternum – mesosternum 12, **42, 44, 52,** 80, 101, 211, 239, 260, 261, 268, 279, 280, 324, 327, 328, 341, 379, 402, 436, **520,** 596, 597, 608, 609, 612, 947
mesostipes – mesostipes 273
mesothorax – mesothorax **51–53,** 206, **505, 508, 517, 520, 523,** 538, 752, 780, 788, 879, 880, 1058
mesovertex – mesovertex 1022
metafurca – metafurca 345

metanotum – metanotum 1, 14, 15, **42, 45, 54, 55,** 70, 82, 90, 103, 104, 109, 118, 178, 183, 211, 250, 260, 262, 280, 285, 295, 322, 329, 337, 342, 346, 359, 381, 383, 392, 393, 403, 404, 411, 416, 434, 435, 465, **518, 521,** 529, 596, 608, 609, 612, 727, 736, 737, 743, 749, 752, 780, 782, 799
metaparameron – metaparamere 284
metapleuron – metapleuron 15, **42, 45,** 82, 83, 183, 259, 260, 262, 279, 280, 381, 465, 492, 608, 609
metasternum – metasternum 15, **42, 45, 55,** 82, 104, 211, 239, 260, 262, 268, 279, 280, 311, 324, 329, 330, 342, 380, 403, 436, 465, 492, **521,** 596, 597, 608, 609, 612, 948
metatarsus – metatarsus **57, 66,** 120, 243, 246, 281, 333, 338, 362, 363, 374–376, 394–397, 493, 494, 496
metathorax – metathorax **54–56,** 206, **505, 508, 518, 521, 524,** 780, 788, 879, 880, 1058
metavertex – metavertex 298, 1022
microseta mesonotalis – microseta of mesonotum 828
– metanotalis – microseta of metanotum 829
mola – molar 112, 370, 574, 685, 795, 825
molle abdominale – abdominal skin 803
mucro – mucro 88, 89
myocicatrix – myocicatrix 729

nesium – nesium 566
notum – notum **46, 47,** 427
– thoracale – thoracic notum 410
– – anterius – anterior thoracic notum 408
– – posterius – posterior thoracic notum 408

occiput – occiput 3, **22–25, 33,** 71, 76, 132, 197, 198, 226, 227, 252, 272, 273, 287, 299, 303, 320, 336, **513,** 546, 554, 556, 648, 671, 728–730, 744, 745, 758, 806, 1021
ocelli glomerati – glomerate eyes 799, 805
ocellus – ocellus **512,** 548, 552, 655, 673, 690, 730, 746, 757, 758, 782
– lateralis – lateral ocellus **21,** 130, 165, 197, 251, 298, 366, 791
– medialis – median ocellus **21,** 130, 165, 197, 251, 298, 366, 791
– primus – first ocellus 903, 904, 911–913
– quartus – fourth ocellus 912
– quintus – fifth ocellus 912
– secundus – second ocellus 912
– tertius – third ocellus 912, 914
ocularium – ocularium 670
oculus – eye **19–24, 32,** 109, 110, 126, 130–135, 160–169, 175, 196, 197, 214, 224, 225, 229–232, 251, 267, 269, 270, 272, 273, 285–288, 297–300, 310, 313, 335, 336, 347–350, 353, 365–368, 384, 385, 405, 454, 458–460, 473, 481–486, 490, 546, 648, 649, 727, 791, 999, 1019, 1021, 1025
omma – omma 763

ommata – ommata 765
orificium vasiforme – vasiform orifice 480
osmeterium – osmeterium 891
ostiolum cerae – ostiole of wax gland 480
– – laterale – lateral ostiole of wax gland 504
– – mediale – median ostiole of wax gland 504
– odoriferum – scent aperture 439

palidium – palidium 614–616
palpifer – palpifer **34,** 226, 231, 234, 255, 271, 273, 368, 369, 547, 560, 597, 671, 728, 806, 813, 930
palpiger – palpiger 5, 7, **37,** 228, 231, 256, 273, 289, 302, 676, 759, 763, 806, 933
palpus – palp 1028, 1029
– labialis – labial palp 3, 5, 7, **22, 24,** 92–94, 110, 114, 115, 132, 139, 140, 195, 198–201, 225–228, 230–232, 236, 252, 253, 256, 270, 272, 273, 286–289, 299, 300, 302, 313, 314, 335, 336, 351, 353, 367, 368, 371, 372, 391, 406, 419, **513–515,** 546, 547, 559–561, 578, 597, 655, 670–673, 676, 677, 728, 729, 745, 758–761, 763–765, 793, 806, 813, 903, 918, 933, 1023, 1029, 1031, 1062
– maxillaris – maxillary palp 1–4, 6, **22, 24,** 72, 73, 77, 94, 95, 109–111, 113, 130–132, 136, 195, 197, 198, 202, 204, 225–227, 230, 232–234, 252, 253, 255, 267, 268, 270–273, 286–289, 299–301, 313, 317, 318, 320, 334–336, 347, 349, 350, 352, 353, 366–369, 389, 390, 419, **513–515,** 546, 547, 551, 552, 554, 558–564, 597, 655, 670–673, 676, 677, 728, 729, 732, 734, 745, 752, 791–794, 806, 813, 903, 930, 1020, 1021, 1023, 1024, 1031, 1058, 1060, 1064
palus – palus 614
papilla – papilla 905
– abdominalis dorsalis – dorsal abdominal papilla 799
– – lateralis – lateral abdominal papilla 799
– analis – anal papilla 1043
– antennalis – antennal papilla 1061
– caudalis dorsalis – dorsal caudal papilla 1041
– – lateralis – lateral caudal papilla 1041
– – ventralis – ventral caudal papilla 1041
– dorsalis – dorsal papilla 1007
– ventralis – ventral papilla 1007
paraclypeus – paraclypeus 730, 744
paraglossa – paraglossa 5, 7, **22, 24, 37, 38,** 93, 114, 198–201, 226–228, 236, 251–253, 256, 272, 273, 287–289, 302, 314, 320, 336, 350, 351, 368, 371–373, 391, **513**
paramentum – paramentum 199
parameron – paramere 284
paranotum – paranotum **46**
– mesothoracale – mesothoracic paranotum 12, 13, **52**
– metathoracale – metathoracic paranotum 15, 16, **55**
– prothoracale – prothoracic paranotum 237, 392
paraproctum – paraproct 161, 217, 219, 382

parasternum abdominale octavum – eighth abdominal parasternite 1068
– – tertium – third abdominal parasternite 1058
– mesothoracale – mesothoracic parasternite 1058
– metathoracale – metathoracic parasternite 1058
paratergum abdominale octavum – eighth abdominal paratergite 1068
– – secundum – second abdominal paratergite 1057
paria – paria 566
paronychium – paronychium 493, 495, 500
pars apicalis mandibulae – apical part of mandible 920
– basalis mandibulae – basal part of mandible 920
paxilla – paxilla 851, 856
pecten analis – anal comb 1044
pedicellus – pedicel **40, 41,** 165, 277, 366, 386, 424, 454, 455, 459, 488, 489, 678, 680, 746, 769, 796, 805, 915, 916
pedium – pedium 566
penicillus – brush of hair 570, 795
– abdominalis – abdominal brush of hair 999
– analis – anal brush of hair 1044
– mesonotalis – mesonotal brush of hair 828
– metanotalis – metanotal brush of hair 829
– praeantennalis – preantennal brush of hair 1022
– pronotalis – pronotal brush of hair 827
– thoracalis – thoracic brush of hair 999
– verticalis exterior – outer vertical brush of hair 1022
– – interior – inner vertical brush of hair 1022
pes – leg **57**
– anterior – fore leg 1, **19, 20,** 70, 90, 109, 126, 160, 161, 195, 196, 224, 250, 267, 268, 285, 297, 313, 334, 346, 365, 383, 404, 408–411, 417, 439, 473, 479, 650
– medialis – middle leg 1, **19, 20,** 70, 90, 109, 126, 160, 161, 195, 196, 224, 250, 267, 268, 285, 297, 313, 334, 346, 365, 383, 404, 408–411, 417, 439, 473, 650
– posterior – hind leg 1, **19, 20,** 70, 90, 109, 126, 160, 161, 195, 196, 224, 250, 267, 268, 285, 297, 313, 334, 346, 365, 383, 404, 408–411, 417, 439, 473, 650
pharynx – pharynx 1028
phoba – phoba 566
pilus branchialis – branchial hair 189
– cercalis – cercal hair 221
– labralis – labral hair 818
pinaculum – pinaculum 905
planta – sole 742, 786, 798, 982
– pulvillata – plantar surface **66,** 470
plegma – plegma 566
plegmatium – plegmatium 566
pleuron abdominale tertium – third abdominal pleurite 527

pleurostoma – pleurostoma **22, 33,** 270, 670, 671, 674, 675, 690, 1059
plica abdominalis intersegmentalis – intersegmental abdominal fold 611, 701
– anuli abdominalis – fold of abdominal ring 799
– branchialis – branchial fold 189
– interanularis quarta – fourth interannular fold 701
– intersegmentalis dorsalis – dorsal intersegmental fold 799
– – ventralis – ventral intersegmental fold 799
– mesothoracalis prima – first mesothoracic fold 599
– – secunda – second mesothoracic fold 599
– metatarsalis – metatarsal fold 362
– metathoracalis intersegmentalis – intersegmental fold of metathorax 600
– – tertia – third metathoracic fold 600
– prothoracalis – prothoracic fold 598
– – intersegmentalis – intersegmental fold of prothorax 598
– – secunda – second prothoracic fold 841
– thoracalis intersegmentalis – intersegmental fold of thorax **525,** 693
pollex – pollex 397, 413, 414
ponticulus hypostomatalis – hypostomal bridge 793
porta epistomatalis – mouth opening 71
porus glandulae – glandular pore 438
– – labialis – glandular pore of labium 371, 671, 672, 677
– mandibularis – mandibular pore 576
postclypeus – postclypeus 225, 269, 313, 366, 367, 481–485, 487, **512,** 670, 730, 744, 790, 792, 903, 1022
postcornu – postcornu 544, 634, 665, 719
postfrons – postfrons **27**
postgena – postgena **32,** 676, 806
postgula – postgula **25,** 76, 729
postlabium – postlabium 3, 5
postmentum – postmentum **24, 37,** 92–94, 114, 140, 164, 169, 170, 175, 198, 199, 287, 289, 350, 351, 368, 391, **514, 515,** 547, 558, 675, 761
postnotum – postnotum **46, 47,** 427
– mesothoracale – mesothoracic postnotum 11, **51, 52,** 80, 117, 183, 260, 357, 780
– metathoracale – metathoracic postnotum 14, **54, 55,** 82, 118, 359, 434, 780
– prothoracale – prothoracic postnotum **48, 49,** 78, 116
postocciput – postocciput **22–25, 33,** 91, 94, 287, 793, 806, 1021
poststernum abdominale – abdominal poststernite 613
– – quartum – fourth abdominal poststernite 18
– mesothoracale – mesothoracic poststernite 104, 105, 117, 209, 240, 294, 358, 399, **523,** 845, 944
– metathoracale – metathoracic poststernite 83, 105, 118, 210, 240, 360, 400, **524,** 781, 846, 945
– prothoracale – prothoracic poststernite 99, 101, 102, 116, 208, 293, 356, 398, 522, 781, 844, 943

posttrochanter – posttrochanter **63,** 84, 735, 847, 848, 850
– anterior – fore posttrochanter 776
– medialis – middle posttrochanter 777
– posterior – hind posttrochanter 778
praeclypeus – preclypeus 225, 269, 670
praefrons – prefrons **27**
praegena – pregena **32**
praegula – pregula **25,** 76
praelabium – prelabium 3, 5, 7
praelabrum – prelabrum 791
praementum – prementum **24, 37,** 92–94, 114, 140, 162–164, 169, 170, 173–175, 198–201, 226, 227, 273, 287, 289, 302, 314, 320, 350, 351, 368, 385, 391, **514, 515,** 547, 558, 561, 597, 648, 675, 676, 728, 729, 745, 761, 768, 806, 813, 1024, 1027
praepleuron metathoracale – metathoracic prepleurite 608, 609
praesternum – presternite **46, 47**
– abdominale – abdominal presternite 107, 613
– – quartum – fourth abdominal presternite 18
– mesothoracale – mesothoracic presternite 12, 13, **53,** 81, 99, 102, 117, 209, 344, **523,** 691, 781, 944
– metathoracale – metathoracic presternite 15, 16, **56,** 83, 105, 118, 210, 345, **524,** 692, 945
– prothoracale – prothoracic presternite 10, **50,** 92, 93, 98, 99, 116, 208, 240, 280, 326, 343, 354, 356, **522,** 781, 943
praetarsus – pretarsus **57, 66,** 120, 246, 263, 281, 333, 338, 362, 363, 374–376, 394–397, 493, 494, 496
praevertex – prevertex **32,** 298, 1022
processus abdominalis – abdominal process 740, 1000
– – lateralis – lateral abdominal process 727
– analis – anal process 1057, 1058, 1066–1068
– coxopleuralis – coxopleural process **58**
– empodialis – empodial digitule 85
– genalis – genal process **32**
– mandibularis – mandibular process 388
– metatarsalis – metatarsal process 376
– palpiformis – palpiform process 92, 94
– papilliformis – papilliform process 1064
– stipitalis – stipital process 233
– thoracalis – thoracic process 1000
– tibialis – tibial process 499
profurca – profurca 343
pronotum – pronotum 1, 8, 9, **42, 43, 48, 49,** 70, 78, 90, 97, 98, 109, 116, 126, 160, 161, 163, 169, 178–181, 183, 195, 196, 205, 211, 214, 224, 230–232, 237–239, 250, 258, 267, 278–280, 285, 293, 297, 306, 310, 311, 313, 322, 323, 325, 334, 337, 340, 346–349, 354, 355, 381, 383, 392, 393, 398, 404, 411, 416, 434, 435, 439, 454, 460–462, 464, 473, 475, 483, 484, 490, 491, **506, 516, 519,**

529, 553, 608–610, 612, 672, 727, 736, 737, 743, 749, 752, 759, 780, 782, 799, 844
propes – proleg **525**
– anterior – fore proleg **505–507,** 529, 635, 655, 672, 677, 690, 727, 736, 743, 752, 788, 789, 799, 879, 943, 946
– medialis – middle proleg **505–507,** 529, 635, 655, 691, 727, 736, 743, 752, 788, 789, 799, 879, 892, 944, 947
– posterior – hind proleg **505–507,** 529, 635, 655, 692, 727, 736, 743, 752, 788, 789, 799, 879, 892, 945, 948
proplegmatium – proplegmatium 566
propleuron – propleuron 9, **42, 43,** 78, 183, 280, 325, 326, 381, 596–598, 608, 609, 672, 673
propodium – abdominal leg **528,** 701–703, 705, 706, 715, 720, 957
– anale – anal leg **507,** 655, 716, 718, 798, 853, 890, 983, 986, 987
– caudale – caudal leg 785
– primum – first abdominal leg 655
– segmenti quarti – abdominal leg of fourth segment **507,** 883, 958
– – quinti – abdominal leg of fifth segment **507,** 895
– – sexti – abdominal leg of sixth segment **507,** 889, 890, 894
– – tertii – abdominal leg of third segment **507**
prosternum – prosternum 9, **42, 43, 49,** 78, 145, 175, 182, 211, 239, 258, 268, 280, 321, 324, 340, 354, 378, 401, 436, 461, 462, 464, **519,** 596, 597, 608, 609, 612, 672, 677, 946
prostheca – prostheca 137, 203, 570, 571, 575, 1030
prothorax – prothorax **48–50, 505, 508, 516, 519, 522,** 538, 546, 673, 677, 760, 780, 788, 879, 880, 1058
protuberantia abdominalis – abdominal protuberance 1038
– – dorsalis – dorsal abdominal protuberance 214
– ocellaris – ocellar protuberance 757
pseudocercus – pseudocercus 283
pseudoculi – pseudoculi 1
pseudoculus – pseudoculus 2
pseudopodium – pseudopod 625
– abdominale – abdominal pseudopod 892, 988, 1011
– thoracale – thoracic pseudopod 988, 1011
pternotorma – pternotorma 566
pulvillus – pulvillus 470, 494, 607
– apicalis – apical pulvillus 363
– subapicalis – subapical pulvillus 363
– tarsalis – tarsal pulvillus 263
pulvinus – pulvinus 385
puparium – puparium **511,** 652–654
pygidium – pygidium 267, 282, 283
pygopodium – pygopod 738, 741, 799, 851–855
pyramis analis – anal pyramid **20,** 160, 187, 312

radicula – radicle **41**
ramus – ramus 89
regio cervicalis – cervical region 306
– genalis – genal region **32,** 71
– gularis – gular region **25**
– occipitalis – occipital region **25**
retinaculum – retinaculum 89, 552, 569, 795, 921
rhinarium – rhinarium 489
rima glossae – glossal excision 173
– ligulae – ligular excision 170
rostellum – rostellum 405
rostrum – rostrum 1–3, 5, 456, 458, 460, 478

scapus – scape **40, 41,** 165, 197, 229–232, 253, 267, 269–272, 274, 298, 310, 366, 384, 386, 419–421, 423, 425, 454, 455, 457, 459, 481, 486, 488, 489, 560, 678–680, 728, 729, 746, 769, 770, 796, 805, 915
scleritum anale – anal sclerite 787
– hypostomatale – hypostomal sclerite 1027–1029
– labiale – labial sclerite 1027
– parastomatale – parastomal sclerite 1028
– pharyngeale – pharyngeal sclerite 1028
– praecoxale – precoxal sclerite 307, 325, 326
– praementi – premental sclerite 675
– subhypostomatale – subhypostomal sclerite 1028
– tarsale – tarsal sclerite 414
scolus abdominalis – abdominal scolus 757, 784, 888, 892
– – tertius – third abdominal scolus 789
– analis – anal scolus 985
– dorsalis – dorsal scolus 985
– capitis – head scolus 892, 939
– caudalis – caudal scolus 892
– lateralis – lateral scolus 985
– mesothoracalis – mesothoracic scolus 757, 789, 941
– metathoracalis – metathoracic scolus 789, 941
– prothoracalis – prothoracic scolus 789
– subspiracularis – subspiracular scolus 969
– supraspiracularis – supraspiracular scolus 970
– thoracalis – thoracic scolus 888, 892
scopa analis – anal fan 1044
– dorsalis – dorsal brush 999, 1044
– femoralis – femoral brush 144
– palpi maxillaris – brush of maxillary palp 4
– – labialis – brush of labial palp 5, 7
– tibialis – tibial brush 144, 748
– ventralis – ventral brush 999, 1044
segmentum abdominale – abdominal segment **67, 68,** 224, **504,** 988

segmentum abdominale decimum – tenth abdominal segment 655, 738, 1066
– – nonum – ninth abdominal segment 757, 799, 983, 1044
– – octavum – eighth abdominal segment 618, 1068
– – primum – first abdominal segment 18, 958
– – quartum – fourth abdominal segment **505, 508**
– – sextum – sixth abdominal segment 161, 383
– – tertium – third abdominal segment 799, 1001
– mesothoracale – mesothoracic segment 637, 1001
– metathoracale – metathoracic segment 637, 1001
– prothoracale – prothoracic segment 637, 1001
– thoracale – thoracic segment **46, 47, 525**
sensillum basiconicum – basiconic sensillum 566, 580, 584, 585, 915
– campaniformium – campaniform sensillum 915, 916
– chaeticulm – sensory bristle 489, 916
– conicum – sensory cone 584
– palpi labiale – sensillum of labial palp 226, 232, 236
– – maxillare – sensillum of maxillary palp 226, 231, 233–235
– styloconicum – styloconic sensillum 915, 919
– trichodeum – sensory hair 488, 580, 583, 915, 916, 930
septula – septula 614, 616
series denticulorum tibialium – row of tibial pegs 309
serra denticuli apicalis – saw of apical denticle 172
seta – bristle 907, 909, 910
– abdominalis – abdominal bristle 437, 635, 1000
– – dorsalis – dorsal abdominal bristle 611, 799
– – dorsolateralis – dorsolateral abdominal bristle 611
– – dorsomedialis – dorsomedial abdominal bristle 611
– – ventralis – ventral abdominal bristle 799
– adfrontalis prima – first adfrontal bristle 902
– – secunda – second adfrontal bristle 902
– analis – anal bristle 851, 852, 855
– – prima – first anal bristle 972
– – quarta – fourth anal bristle 972
– – secunda – second anal bristle 972
– – tertia – third anal bristle 972
– antennalis – antennal bristle **41, 796**
– – apicalis – apical antennal bristle 746, 805
– – terminalis – terminal antennal bristle 1061
– anterior prima – first anterior bristle 901, 911
– – secunda – second anterior bristle 901, 911
– – tertia – third anterior bristle 904, 911
– branchialis – branchial bristle 189
– capitata – capitate bristle 757
– cardinalis – cardinal bristle 813, 1064
– caudalis – caudal bristle 635
– cercalis – cercal bristle 221

seta clavata – clavate bristle 85, 797
- clypealis – clypeal bristle 1019, 1059
- – exterior – outer clypeal bristle 1021
- – interior – inner clypeal bristle 1021
- – posterior – posterior clypeal bristle 1021
- – prima – first clypeal bristle 902
- – secunda – second clypeal bristle 902
- coxalis apicalis – apical coxal bristle 847
- dorsalis – dorsal bristle 983, 1035
- – prima – first dorsal bristle 954–958
- – quarta – fourth dorsal bristle 954–956
- – quinta – fifth dorsal bristle 954
- – secunda – second dorsal bristle 954–958
- – tertia – third dorsal bristle 954–956
- dorsolateralis – dorsolateral bristle 1035
- epicranialis – epicranial bristle 792
- – octava – eighth epicranial bristle 595
- – prima – first epicranial bristle 595
- – quarta – fourth epicranial bristle 595
- – quinta – fifth epicranial bristle 595
- – secunda – second epicranial bristle 595
- – septima – seventh epicranial bristle 595
- – sexta – sixth epicranial bristle 595
- – tertia – third epicranial bristle 595
- epipleuralis – epipleural bristle 954
- epistomatalis – epistomal bristle 1019
- femoralis – femoral bristle 735, 950
- – apicalis – apical femoral bristle 847
- frontalis – frontal bristle 2, 794, 901, 1059
- – exterior – outer frontal bristle 1021
- – interior – inner frontal bristle 1021
- – medialis – median frontal bristle 1021
- – prima – first frontal bristle 595, 1019
- – quarta – fourth frontal bristle 595
- – quinta – fifth frontal bristle 595
- – secunda – second frontal bristle 595, 1019
- – tertia – third frontal bristle 595
- frontoclypealis anterior prima – first anterior frontoclypeal bristle 805
- – lateralis – lateral frontoclypeal bristle 805
- – posterior – posterior frontoclypeal bristle 805
- galealis – galeal bristle **35**
- genalis – genal bristle 674, 1021
- glossalis – glossal bristle **39**, 373
- hypopleuralis – hypopleural bristle 954
- hypopodii – bristle of hypopodite 972
- labialis – labial bristle 1062

seta labralis – labral bristle 1059, 1065
– – apicalis – apical labral bristle 818
– – basalis – basal labral bristle 818
– – lateralis prima – first lateral labral bristle 917
– – – secunda – second lateral labral bristle 917
– – – tertia – third lateral labral bristle 917
– – medialis – median labral bristle 818
– – – prima – first median labral bristle 917
– – – secunda – second median labral bristle 917
– – – tertia – third median labral bristle 917
– – prima – first labral bristle 1019
– – secunda – second labral bristle 1019
– – tertia – third labral bristle 1019
– lacinialis – lacinial bristle **36**
– lateralis – lateral bristle 902, 904, 911, 958, 1035
– – prima – first lateral bristle 901, 956
– – secunda – second lateral bristle 901, 956
– – tertia – third lateral bristle 956
– lobi lateralis – bristle of lateral lobe 171, 172
– mandibularis – mandibular bristle **26**, 747, 792, 795, 823, 920, 1030, 1063
– marginalis – marginal bristle 1068
– maxillaris – maxillary bristle 1060
– medialis – median bristle 1035
– mentalis – mental bristle 371
– mesonotalis anterior – anterior bristle of mesonotum 828
– – lateralis – lateral bristle of mesonotum 610, 828
– – medialis – median bristle of mesonotum 610, 828
– – posterior – posterior bristle of mesonotum 610, 828
– mesosternalis – mesosternal bristle 399
– metanotalis anterior – anterior bristle of metanotum 829
– – lateralis – lateral bristle of metanotum 829
– metasternalis – metasternal bristle 400, 403
– occipitalis exterior – outer occipital bristle 1021
– – interior – inner occipital bristle 1021
– ocellaris – ocellar bristle 746
– – prima – first ocellar bristle 911
– – secunda – second ocellar bristle 911
– – tertia – third ocellar bristle 911, 913
– orbicularis – orbicular bristle 805
– palpi labialis – bristle of labial palp 7, 372, 932
– paraglossalis – paraglossal bristle **38**
– parasternalis – parasternal bristle 1058
– paratergalis – paratergal bristle 1057
– pedicellaris – pedicellar bristle **41**
– pleurostomatalis – pleurostomal bristle 1059
– plumosa – plumose bristle 85, 797

seta posterior prima – first posterior bristle 901, 902, 904
– – secunda – second posterior bristle 901, 902, 904
– praementalis – premental bristle 170
– praetarsalis – pretarsal bristle 377, 775
– pronotalis – pronotal bristle 596
– – anterior – anterior bristle of pronotum 610, 826
– – lateralis – lateral bristle of pronotum 610
– – medialis – median bristle of pronotum 610, 826
– – posterior – posterior bristle of pronotum 610, 827
– – – prima – first posterior bristle of pronotum 826
– – – secunda – second posterior bristle of pronotum 826
– – posterolateralis – posterolateral bristle of pronotum 827
– propodii – bristle of abdominal leg 957, 958, 977
– prosternalis – prosternal bristle 398
– pygopodii – bristle of pygopod 852
– – dorsalis – dorsal bristle of pygopod 854
– – ventralis – ventral bristle of pygopod 854
– scapalis – bristle of scape **41**
– simplex – simple bristle 85
– spathulata – spathulate bristle 437
– spatulata – spatulate bristle 488, 502, 772
– spiracularis prima – first spiracular bristle 957, 958
– – quarta – fourth spiracular bristle 957, 958
– – secunda – second spiracular bristle 957
– – tertia – third spiracular bristle 957
– sternalis anterior – anterior sternal bristle 1058
– – posterior – posterior sternal bristle 1058
– stipitalis – stipital bristle 813, 1064
– submentalis – submental bristle 813
– tergalis – tergal bristle 851, 853
– – anterior – anterior tergal bristle 1057
– – posterior – posterior tergal bristle 1057
– tarsalis – tarsal bristle 339, 412, 429, 495, 500, 502, 774, 950
– – lanceolata – lanceolate tarsal bristle 952
– terminalis – terminal bristle 799
– tibialis – tibial bristle **65**, 376, 412, 735, 950
– – apicalis – apical tibial bristle 847
– tibiotarsalis – tibiotarsal bristle 949
– uropodii – bristle of uropodite 972
– ventralis – ventral bristle 977, 1035
– – prima – first ventral bristle 956
– – quarta – fourth ventral bristle 954, 956
– – secunda – second ventral bristle 954, 956
– – tertia – third ventral bristle 956, 957
– ventrolateralis – ventrolateral bristle 1035
– verticalis – vertical bristle 2, 794

seta verticalis anterior prima – first anterior vertical bristle 805
– – dorsalis prima – first dorsal vertical bristle 1059
– – – secunda – second dorsal vertical bristle 1059
– – – tertia – third dorsal vertical bristle 1059
– – lateralis prima – first lateral vertical bristle 1060
– – – secunda – second lateral vertical bristle 1060
– – posterior secunda – second posterior vertical bristle 805
– – prima – first vertical bristle 901, 1019
– – secunda – second vertical bristle 901, 1019
– – tertia – third vertical bristle 901
– – ventralis – ventral vertical bristle 1060
setula dorsalis – dorsal setule 958
– frontalis – frontal setule 1059
– lateralis – lateral setule 958
– ventralis – ventral setule 958
– verticalis anterior – anterior vertical setule 1059
– – posterior – posterior vertical setule 1059
sinus unci – sinus of uncus 855
– verticalis coronalis – coronal sinus of vertex 902
sipho – syphon 422, 453
siphunculus – siphuncle 476
spatula – spatula 982
spina – spine 908
– abdominalis – abdominal spine 474, 799, 1038
– – dorsalis – dorsal abdominal spine 160, 186, 475, 885
– – lateralis – lateral abdominal spine 160, 186
– analis – anal spine 787
– antennalis – antennal spine 1061
– branchialis – branchial spine 189
– discoidalis – discoidal spine 308, 309
– dorsalis – dorsal spine 958, 985
– labralis – labral spine 818, 1065
– mandibularis – mandibular lancet 419–421, 456, 765
– maxillaris – maxillary lancet 420, 421, 456, 765
– maxillomandibularis – maxillomandibular lancet 761, 762
– metanotalis – metanotal spine 393
– – lateralis – lateral metanotal spine 392
– metatarsalis – metatarsal spine 376
– pollicis – spine of pollex 414
– pronotalis – pronotal spine 393
– – lateralis – lateral pronotal spine 392
– – medialis – median pronotal spine 392
– styli – spine of stylus 124
– subspiracularis – subspiracular spine 958
– supraspiracularis – supraspiracular spine 958
– tibialis – tibial spine **65,** 241, 362, 472, 493

spina tibialis digitiformis – digitiform spine of tibia 245
- tibiotarsalis – tibiotarsal spine 242
- unguis – spine of claw 377
- ventralis – ventral spine 958
spinasternum – spinasternum **46**
- mesothoracale – mesothoracic spinasternum 311
- metathoracale – metathoracic spinasternum 16, 83
- prothoracale – prothoracic spinasternum 79, 237, 257
spiraculum – spiracle 79, 98, 101, 104, 125, 258, 283, 306, 325, 327, 329, 381, 465, 483, 598, 610, 618, 625, 627, 707, 782
- abdominale – abdominal spiracle **69**, 247, 296, 409, 415, 438, **527**, 529, 701, 715, 957
- – octavum – eighth abdominal spiracle 629, 717
- – primum – first abdominal spiracle 596, 958
- – quintum – fifth abdominal spiracle 789
- – septimum – seventh abdominal spiracle **507**, 630, 655
- – sextum – sixth abdominal spiracle 879, 898
- – tertium – third abdominal spiracle **506**, 750
- anterius – anterior spiracle 1001
- caudale – caudal spiracle 1007, 1040, 1041
- mesothoracale – mesothoracic spiracle 11, **52**, 293
- metathoracale – metathoracic spiracle 14, **55**, 294
- posterius – posterior spiracle 1001
- prothoracale – prothoracic spiracle 655, 672, 673, 879, 954, 1007
- thoracale – thoracic spiracle 247, 529, 596, 693
stemmata – stemmata 744, 745, 792
sterna abdominalia – abdominal sternites **69**
sternellum – sternellum **47**
- mesothoracale – mesothoracic sternellum **53**
- metathoracale – metathoracic sternellum **56**
- prothoracale – prothoracic sternellum **50**
sternopleuron – sternopleuron **47**, **67**, 121
- abdominale – abdominal sternopleuron 613
- dorsale – dorsal sternopleuron **68**
- mesothoracale – mesothoracic sternopleuron **53**, 182, 209, 240, 327, 328, 341, 358, 691
- metathoracale – metathoracic sternopleuron **56**, 118, 182, 210, 240, 329, 330, 342, 360, 403, 692
- prothoracale – prothoracic sternopleuron **50**, 99, 116, 182, 208, 237, 240, 354, 690
- ventrale – ventral sternopleuron **68**
sternum – sternite **67**, **68**, 427
- abdominale – abdominal sternite 107, 196, 226, 403, 613
- – decimum – tenth abdominal sternite 125, 248, 282, 715
- – nonum – ninth abdominal sternite 248, 283, 716, 785
- – octavum – eighth abdominal sternite 219, 283, 503, 784, 1068
- – penultimum – penultimate abdominal sternite 268, 284

sternum abdominale primum – first abdominal sternite 82, 211, 295, 296, 342, 380, 464, 465
- – quartum – fourth abdominal sternite 123
- – quintum – fifth abdominal sternite **527**
- – secundum – second abdominal sternite 216
- – septimum – seventh abdominal sternite 18, 285, 364, 503, 618
- – sextum – sixth abdominal sternite 108
- – tertium – third abdominal sternite 123, 324, 382, 783, 1058
- caudale – caudal sternite 265
- decimum – tenth sternite **69**
- mesothoracale – mesothoracic sternum 1058
- metathoracale – metathoracic sternum 1058
- nonum – ninth sternite **69**
- octavum – eighth sternite **69**
- primum – first sternite **69**
- prothoracale – prothoracic sternum 1058
- quartum – fourth sternitte **69**
- quintum – fifth sternite **69**
- secundum – second sternite **69**
- septimum – seventh sternite **69**
- sextum – sixth sternite **69**
- tertium – third sternite **69**
- thoracale – thoracic sternum 407
stigma – stigma 640
stipes – stipes 4, 6, **22, 24, 34,** 77, 95, 111, 113, 136, 177, 198, 202, 204, 226, 227, 233–235, 252, 253, 255, 272, 286, 289, 299, 301, 318–320, 336, 350, 352, 368, 369, 389, 390, **514, 515,** 547, 560, 561, 563–565, 597, 677, 728–730, 732, 734, 745, 759, 761, 763, 765, 768, 806, 813, 903, 931, 1020, 1023, 1024, 1064
stylus – style 107, 108, 123, 125, 248, 249, 477
- primus – first style 18
- secundus – second style 18
- tertius – third style 18
subcoxa – subcoxa **61,** 693, 697
submentum – submentum **23,** 200, 226, 227, 299, 302, 560, 561, 597, 648, 676, 728, 745, 806, 813, 903, 1023
subtrochanter – subtrochanter **63,** 84, 735, 847, 848, 850
- anterior – fore subtrochanter 776
- medialis – middle subtrochanter 7677
- posterior – hind subtrochanter 778
sulcus frontalis – frontal groove 648
- mediocranialis – mediocranial groove 648
- verticalis – vertical groove 648
superlingua – superlingua 138
sutura acrotergalis mesothoracalis – acrotergal suture of mesothorax **51**
- – metathoracalis – acrotergal suture of metathorax **54**
- – prothoracalis – acrotergal suture of prothorax **48**

sutura adfrontalis – adfrontal suture **512, 513,** 790, 902
- antennalis – antennal suture **41**
- basicostalis – basicostal suture **59–61**
- basicoxalis – basicoxal suture **61**
- branchialis – branchial suture 189
- cardinostipitalis – cardinostipital suture **34**
- cephalica – cephalic suture 1058, 1060
- circumantennalis – circumantennal suture 269, 271
- clypeofrontalis – clypeofrontal suture 269
- clypeogenalis – clypeogenal suture **21, 27–30, 33,** 384, 763, 791
- clypeolabralis – clypeolabral suture **21, 28, 31, 33,** 286, **512,** 560, 670
- coronalis – coronal suture **21, 23, 27, 28,** 131, 133, 165, 225, 269, 286, 298, 335, 347, **512,** 548, 595, 670, 676, 730, 744, 759, 760, 790, 791, 805, 806, 1019, 1022, 1059
- – anterior – anterior coronal suture 792
- – posterior – posterior coronal suture 793
- coxalis – coxal suture **58, 60, 61,** 121, **526,** 951
- epicranialis – epicranial suture 91, 251, 313, 320
- epistomatalis – epistomal suture 271, 730, 792
- frontalis – frontal suture 72, 225, 251, 269, 271, 272, 298, 335, **512, 513,** 549, 560, 670, 690, 730, 744, 790, 792, 902, 1059
- – lateralis – lateral frontal suture **27**
- – transversalis – transversal frontal suture **27, 28**
- frontoclypealis – frontoclypeal suture **21, 27, 29, 30,** 348, **512,** 560, 670, 672
- frontogenalis – frontogenal suture 384
- frontoverticalis – frontovertical suture **21,** 805, 1019, 1022
- gularis – gular suture 350, **514,** 728, 729, 745, 806, 816
- – epistomatalis – epistomal suture of gula **25**
- – lateralis – lateral gular suture **25**
- hypostomatalis – hypostomal suture 555
- labialis – labial suture **37**
- mesonotalis – mesonotal suture **51,** 80
- – medialis – median suture of mesonotum 11
- mesopleuralis – mesopleural suture **52,** 294, 327, 341, 379, 464, **520**
- mesosternalis – mesosternal suture 436
- metanotalis – metanotal suture **54**
- – medialis – median suture of metanotum 14
- metapleuralis – metapleural suture **55,** 247, 295, 329, 342, 464, **521**
- occipitalis – occipital suture 3, **22, 23, 25, 32, 33,** 71, 72, 226, 227, 252, 271, 273, 287, 299, 336, 554, 676, 728, 730, 744, 806
- ocularis – ocular suture 32
- pleuralis – pleural suture **47, 58, 59,** 332, **525**
- pleurostomatalis – pleurostomal suture 270
- postfrontalis – postfrontal suture **21, 22,** 91, 94, 110, 347, 348, 353
- postgenalis – postgenal suture 676
- postoccipitalis – postoccipital suture **22, 23, 25,** 91, 94, 806

sutura postorbitalis – postorbital suture 730
- praesternalis – presternal suture **46, 47**
- – mesothoracalis – presternal suture of mesothorax 13, **53,** 81
- – metathoracalis – presternal suture of metathorax 16, **56,** 83
- – prothoracalis – presternal suture of prothorax 10, **50**
- pronotalis – pronotal suture **48,** 78
- – transversalis – transversal suture of pronotum 305, 306
- propleuralis – propleural suture **49,** 321, 325, 462, **519**
- sternacostalis – sternacostal suture **47**
- – mesothoracalis – sternacostal suture of mesothorax **53**
- – metathoracalis – sternacostal suture of metathorax 16, **56**
- – prothoracalis – sternacostal suture of prothorax **50,** 79
- sternopleuralis prothoracalis – sternopleural suture of prothorax **50**
- subantennalis – subantennal suture **21, 22, 32,** 229, 335
- subcostalis – subcostal suture **61**
- subcoxalis – subcoxal suture **61**
- subgenalis – subgenal suture **32, 33**
- submarginalis mesothoracalis – submarginal suture of mesothorax **523**
- – metathoracalis – submarginal suture of metathorax **524**
- – prothoracalis – submarginal suture of prothorax **522**
- subocellaris – subocellar suture 71, 72
- subocularis – subocular suture **22, 32,** 288, 335
- transversalis – transversal suture 299
- trochanteralis – trochanteral suture **63**
- verticalis — vertical suture 251
- – transversalis – transversal suture of vertex 21

tarsomeron mediale – median subjoint of tarsus **57, 66,** 362, 363, 494
- secundum – second subjoint of tarsus 333
- tertium – third subjoint of tarsus 333
tarsungulus – tarsungulus 603, 604
tarsus – tarsus 17, **57, 66,** 106, 119, 120, 122, 144, 215, 243–245, 308, 331, 412–414, 428, 466, 467, 471, 472, 495, 497–499, 501, **525, 526,** 601, 605, 693, 694, 697, 735, 748, 774, 847, 848, 850, 950–953
- anterior – fore tarsus **42,** 141, 290, 598, 749, 776
- medialis – middle tarsus **42,** 142, 291, 599, 777
- posterior – hind tarsus **42,** 143, 292, 600
teges – teges 616, 617
tegillum – tegillum 614
tegula – tegula 361
telson – telson 18, 195, 538
tempus – temple **22–24, 32,** 132, 229, 231, 232, 253, 287, 288, 299, 300, 367, 368, 670, 671, 676
tenaculum – clasp 395
tentorium – tentorium 421
terga abdominalia – abdominal tergites **69,** 346

tergum – tergite **67, 68**
- abdominale – abdominal tergite 147, 150, 196
- – decimum – tenth abdominal tergite 125, 185, 219, 248, 282, 296, 783, 853
- – nonum – ninth abdominal tergite 219, 249, 266, 283, 437, 438, 738, 851, 853, 855
- – octavum – eighth abdominal tergite 160, 503, 623, 741, 1068
- – primum – first abdominal tergite 82, 206, 216, 247, 278, 296, 322, 342, 359, **527**, 750, 1057
- – quartum – fourth abdominal tergite 1, 18, 285, 382, 529
- – quintum – fifth abdominal tergite 750
- – secundum – second abdominal tergite 123
- – tertium – third abdominal tergite 89, 108, 415
- – ultimum – last abdominal tergite 267
- caudale – caudal tergite 265
- decimum – tenth tergite **69**
- mesothoracale – mesothoracic tergite 1057
- metathoracale – metathoracic tergite 1057
- nonum – ninth tergite **69**
- octavum – eighth tergite **69**
- primum – first tergite **69**
- prothoracale – prothoracic tergite 1057, 1058
- quartum – fourth tergite **69**
- quintum – fifth tergite **69**
- secundum – second tergite **69**
- septimum – seventh tergite **69**
- sextum – sixth tergite **69**
- tertium – third tergite **69**

tergopleuron – tergopleuron **67**
- abdomine octavum – eighth abdominal tergopleuron 364
- – secundum – second abdominal tergopleuron 415
- dorsale – dorsal tergopleuron **68**
- mesothoracale – mesothoracic tergopleuron 379, 381, 608, 609
- metathoracale – metathoracic tergopleuron 381
- – anterius – anterior tergopleuron of metathorax 608, 609
- – posterius – posterior tergopleuron of metathorax 608, 609
- ventrale – ventral tergopleuron **68**

theca abdominalis – abdominal cover **509, 510,** 654, 751, 870, 1069
- alae – wing cover 510
- – anterioris – fore wing cover **19, 20,** 126, 160, 161, 179, 180, 195, 196, 205, 207, 216, 224, 238, 297, 306, 310, 313, 323, 334, 361, 365, 417, 418, 439, 464, 473, 474, 490–492, **509,** 751, 756, 870
- – posterioris – hind wing cover **19, 20,** 126, 160, 161, 179, 180, 195, 196, 205, 207, 224, 264, 267, 278, 297, 313, 323, 334, 361, 365, 417, 418, 464, 473, 490, **509,** 649, 654, 870
- antennalis – antennal cover 418, **509, 510,** 751, 870
- capitis – head cover **509, 510,** 654, 751, 870, 1069

theca mesothoracalis – mesothoracic cover 870
– metathoracalis – metathoracic cover 870
– prothoracalis – prothoracic cover 870
– tegminis – tegminal cover 264, 267, 279
– thoracalis – thoracic cover **509**, 654, 1069
thorax – thorax 1, **19, 20, 42,** 70, 90, 109, 126, 160, 195, 224, 250, 285, 297, 313, 334, 346, 365, 383, 404, 418, 439, 473, **505–508,** 529, 635, 636, 642, 655, 727, 743, 752, 788, 880, 999, 1057
tibia – tibia 17, **57, 65,** 106, 119, 120, 144, 215, 241, 243–245, 281, 308, 309, 331, 338, 363, 374–376, 394–397, 412–414, 428, 466–468, 471, 472, 493, 496–499, 501, **525, 526,** 601, 603–605, 693, 694, 697, 735, 748, 847, 848, 850, 950, 951
– anterior – fore tibia **42,** 141, 290, 598, 776
– medialis – middle tibia **42,** 142, 291, 599, 749, 777
– posterior – hind tibia **42,** 143, 292, 600, 778
tibiotarsus – tibiotarsus 84, 85, 242, 469, 602, 695, 772, 773, 949
trachea – trachea 148
– branchialis – branchial trachea 189
tracheola subcostalis – subcostal tracheole 189
– terminalis – terminal treacheole 189
trochanter – trochanter 17, **57,** 106, 119, 120, 144, 215, 241, 243, 245, 281, 308, 309, 331, 338, 374, 375, 394, 396, 428, 466, 467, 498, **525, 526,** 601–605, 693, 697, 748
– anterior – fore trochanter **42,** 290, 356, 462
– compositus – posttrochanter et subtrochanter **63**
– medialis – middle trochanter **42,** 259, 291, 324, 463, 491, 599
– posterior – hind trochanter **42,** 259, 292, 465, 492
– simplex – simple trochanter **62**
trochantinus – trochantin **46, 58, 59,** 121, 332
– anterior – fore trochantin **49,** 78, 98, 116, 293, 307, 321, 325, 326, 340, 343, 354
– medialis – middle trochantin **52,** 80, 101, 117, 294, 327, 328, 341, 344, 358, 379, 612
– posterior – hind trochantin **55,** 82, 104, 105, 118, 295, 329, 330, 342, 345, 360, 380, 612
truncus trachealis – tracheal trunk 189
tuba analis – anal tube 475, 787
tuberculum abdominale dorsale – dorsal abdominal tubercle 752
– – laterale – lateral abdominal tubercle 752
– antennale – antennal tubercle 454, 483
– dorsale – dorsal tubercle 1040
– laterale – lateral tubercle 1040
– mesothoracale – mesothoracic tubercle 752
– metathoracale – metathoracic tubercle 752
– prothoracale – prothoracic tubercle 752
– ventrale – ventral tubercle 1040
tubulus analis – anal tubulus 988, 998

tubus spiracularis – respiratory siphon 999, 1010, 1011 1044, 1052
– ventralis – ventral tube 82
tympanum – auditory organ 247
– tibiale – auditory organ of tibia 243

uncus analis – anal hook 799, 851–855, 858
– pseudopodii – hook of pseudopod 1036
unguis – claw 17, **57, 66,** 84–86, 106, 119, 120, 122, 142, 144, 215, 241–246,
    263, 281, 308, 331, 333, 338, 339, 362, 363, 374–376, 394–397, 412–414,
    428–433, 466, 467, 470, 472, 493–502, **525, 526,** 601, 602, 605, 673, 693, 696,
    699, 700, 735, 748, 772–775, 779, 847–849, 949–953
– tarsi anterioris – claw of fore tarsus 290, 776
– – medialis – claw of middle tarsus 291, 777
– – posterioris – claw of hind tarsus 292, 778
urogomphus – urogomphus 621–625
uropodium – uropodite **528,** 701, 972

valva – valve 503
– analis – anal valve 125
valvula anterior – anterior valvula 249
– dorsalis – dorsal valvula 249
– posterior – posterior valvula 249
ventrocoxa – ventrocoxa **60, 61, 526,** 951
verruca dorsalis – dorsal verruca 968
– subspiracularis – subspiracular verruca 968
– supraspiracularis – supraspiracular verruca 968
– ventralis – ventral verruca 968
vertex – vertex 1–4, **21–23,** 71–73, 110, 132, 135, 162, 163, 165, 167–169, 197,
    225–227, 229–231, 269–273, 286, 288, 299, 300, 335, 336, 347, 366–368, 419,
    420, 458, 460, 481–483, 485, 486, **513,** 546, 550, 560, 648, 670–673, 675, 676,
    690, 730, 744, 745, 758–761, 765, 790, 791, 793, 794, 805, 806, 901, 1019,
    1021, 1059, 1060
vesicula – vesicle 87, **528,** 739, 742, 786, 797, 798, 972
– caudalis – caudal vesicle 750
– coxalis – coxal vesicle 123

zygum – zygum 566

ENGLISH–LATIN

The bold face numbers refer to figures in the two general parts.

abdomen – abdomen 1, **19, 20, 69,** 70, 90, 109, 126, 160, 195, 224, 250, 267, 285, 297, 313, 334, 346, 365, 383, 404, 418, 439, 473, **505–508, 527,** 529, 635, 636, 642, 655, 727, 743, 752, 788, 880, 1057
abdominal appendage – appendix abdominalis 107, 108
– bristle – seta abdominalis 434, 635, 1000
– brush of hair – penicillus abdominalis 999
– cover – theca abdominalis **509, 510,** 654, 751, 870, 1069
– coxosternite – coxosternum abdominale 123, 248, 266
– episternum – episternum abdominale 613
– filament – filamentum abdominale 742, 882
– hemisternite – hemisternum abdominale 364
– hemitergite – hemitergum abdominale 346
– hyposternum – hyposternum abdominale 613
– laterotergite – laterotergum abdominale 296
– leg – propodium **528,** 701–703, 705, 706, 715, 720, 799, 957
– – of fifth segment – propodium segmenti quinti **507,** 895
– – – fourth segment – propodium segmenti quarti **507,** 883, 958
– – – sixth segment – propodium segmenti sexti **507,** 889, 890, 894
– – – third segment – propodium segmenti tertii **507**
– line – linea abdominalis **69, 506**
– membrane – membrana abdominalis 382
– – of fourth segment – membrana abdominalis quarta 18
– poststernite – poststernum abdominale 613
– presternite – praesternum abdominale 107, 613
– process – processus abdominalis 740, 1000
– protuberance – protuberantia abdominalis 1038
– pseudopod – pseudopodium abdominale 892, 988, 1011
– ring – anulus abdominalis **506,** 618, 783, 799, 1007
– scolus – scolus abdominalis 757, 784, 888, 892
– segment – segmentum abdominale **67, 68,** 224, **504,** 988
– skin – coriaceum abdominale 1057, 1058
– skin – membranaceum abdominale **506,** 799
– skin – molle abdominale 803
– spine – spina abdominalis 474, 799, 1038

abdominal spiracle – spiraculum abdominale **69**, 247, 296, 409, 415, 438, **527**, 529, 701, 715, 957
– sternite – sternum abdominale 107, 196, 226, 403, 613
– sternites – sterna abdominalia **69**
– sternopleuron – sternopleuron abdominale 613
– tergite – tergum abdominale 147, 150, 196
– tergites – terga abdominalia **69**, 346
– tracheal gill – branchia abdominalis 150, 196, 540, 727, 799
acanthoparia – acanthoparia 566
acroparia – acroparia 566
acrotergal suture of mesothorax – sutura acrotergalis mesothoracalis **51**
– – – metathorax – sutura acrotergalis metathoracalis **54**
– – – prothorax – sutura acrotergalis prothoracalis **48**
acrotergite – acrotergum 427
addorsal line – linea addorsalis 878, 880
adfrons – adfrons **512, 513,** 790, 901, 902
adfrontal suture – sutura adfrontalis **512, 513,** 790, 902
admentum – admentum 92, 93
alimentary canal – canalis alimentarius 762
alveole – alveolus **41**
ampulla of second segment – ampulla segmenti secundi **507**
– – seventh segment – ampulla segmenti septimi **507**
anal appendage – appendicula analis 620
– area – area analis 1040
– bristle – seta analis 851, 852, 855
– brush of hair – penicillus analis 1044
– comb – ctenidium anale 983, 1057, 1066, 1067
– comb – pecten analis 1044
– disc – discus analis 504
– fan – scopa analis 1044
– hook – uncus analis 799, 851–855, 858
– horn – cornu anale 984
– leg – propodium anale **507,** 655, 716, 718, 798, 853, 890, 983, 986, 987
– lobe – lobus analis 623, 1040
– papilla – papilla analis 1043
– plate – lamina analis 972
– process – processus analis 1057, 1058, 1066–1068
– pyramid – pyramis analis **20,** 160, 187, 312
– sclerite – scleritum anale 787
– scolus – scolus analis 985
– spine – spina analis 787
– tracheal gill – branchia analis 628, 799, 853, 854, 999, 1011, 1042, 1044
– tube – tuba analis 475, 787
– tubulus – tubulus analis 988, 998
– valve – valva analis 125
anapleurite – anapleuron **46,** 525

anepimeral sclerite – anepimeron 427
– – of mesothorax – anepimeron mesothoracale 434, 435
– – – metathorax – anepimeron metathoracale 434, 435
– – – prothorax – anepimeron prothoracale 279, 435, 612
anepisternal sclerite – anepisternum 47, 427
– – of prothorax – anepisternum prothoracale 49, 279, 776
annular joint of antenna – articulus antennalis anularis 40
anteclypeus – anteclypeus 313, 366, 367, 481–485, 487, **512**, 730, 744, 790, 792, 903, 1022
antenna – antenna **19–22, 40,** 70–73, 76, 90–92, 94, 109, 110, 126, 130–133, 135, 160–166, 169, 175, 195–198, 214, 224, 225, 250, 251, 267, 268, 285, 286, 288, 297, 300, 313, 334, 335, 346–350, 365, 367, 383–385, 387, 404, 405, 416, 417, 439, 473, 477, 478, 482, 485, 490, **505, 512, 513, 515,** 547, 551–553, 555–560, 597, 648, 655, 671–675, 677, 690, 727–730, 743–745, 752, 757, 758, 760, 763–765, 782, 788, 790–792, 794, 901, 903, 904, 943, 946, 988, 999, 1020–1022, 1025–1029, 1031, 1034, 1057, 1059
antennal appendage – lamina antennalis 40
– area – area antennalis 796, 1061
– areola – areola antennalis 426
– bristle – seta antennalis **41,** 796
– cover – theca antennalis 418, **509, 510,** 751, 870
– furrow – fossa antennae 387
– joint – articulus antennalis **41,** 488, 679, 916, 1061
– lobe – lobus antennalis 40
– membrane – corium antennale **41,** 225, 269, 271, 546, 587, 678, 902, 903, 1025
– mount – colliculus antennalis 1059, 1061
– papilla – papilla antennalis 1061
– sclerite – antennarium 670, 672, 678, 679
– socket – cavea antennalis **21, 32,** 251, 298, 347, 353, 366, 481, 791, 799, 805
– spine – spina antennalis 1061
– suture – sutura antennalis 41
– tubercle – tuberculum antennale 454, 483
– antennifer – antennifer **32,** 271, 581
anterior → see also first, fore
– alveole – alveolus anterior 904
– bristle of mesonotum – seta mesonotalis anterior 828
– – – metanotum – seta metanotalis anterior 829
– – – pronotum – seta pronotalis anterior 610, 826
– condyle of mandible – condylus mandibularis anterior **33**
– coronal suture – sutura coronalis anterior 792
– lateral plate of mesonotum – lamina mesonotalis lateralis anterior 840
– laterocervical sclerite – lamina laterocervicalis antderior 303
– line of pronotum – linea pronotalis anterior 305
– lobe of pronotum – lobus pronotalis anterior 169
– – – trochanter – lobus trochanteralis anterior **62, 63**
– mediocervical sclerite – lamina mediocervicalis anterior 303

anterior mediolateral plate of mesonotum – lamina mesonotalis mediolateralis anterior 840
– – – – metanotum – lamina metanotalis mediolateralis anterior 840
– ocular margin – clavus ocularis anterior 21
– spiracle – spiraculum anterius 1001
– sternal bristle – seta sternalis anterior 1058
– tentorial pit – fovea tentorialis anterior 91, 648, 675, 730, 792
– tergal bristle – seta tdergalis anterior 1057
– tergopleuron of metathorax – tergopleuron metathoracale anterius 608, 609
– thoracic notum – notum thoracale anterius 408
– valvula – valvula anterior 249
– vertical setule – setula verticalis anterior 1059
anus – anus 184, 504, 614, 618, 715, 717, 720, 983, 1042, 1043
apex of antenna – apex antennalis 40, 386, 457, 679
– – claw – apex unguis 85
– – galea – apex galealis 35
– – glossa – apex glossalis 39
– – labial palp – apex palpi labialis 37
– – lacinia – apex lacinialis 36
– – lower molar lobe – apex lobi molaris inferior 26
– – mandible – apex mandibularis 26, 134, 388, 824
– – maxillary palp – apex palpi maxillaris 34, 255
– – paraglossa – apex paraglossalis 38
– – tracheal gill – apex branchialis 148, 189
– – upper molar lobe – apex lobi molaris superior 26
apical antennal bristle – seta antennalis apicalis 746, 805
– cervical sclerite – lamina cervicalis apicalis 293
– condyle of mandible – condylus mandibularis apicalis 388
– coxal bristle – seta coxalis apicalis 847
– – cavity – cavea coxalis apicalis 60, 61
– denticle – denticulus apicalis 88, 170, 171, 173
– – of lacinia – denticulus laciniae apicalis 235, 318
– femoral bristle – seta femoralis apicalis 847
– – cavity – cavea femoralis apicalis 64
– joint – artus apicalis 797
– – of antenna – articulus antennalis apicalis 40, 489, 579, 678, 746
– – – cercus – articulus cercalis apicalis 364
– – – labial palp – artus palpi labialis apicalis 37, 771, 932
– – – maxillary palp – artus palpi maxillaris apicalis 34, 77, 931
– – – rostrum – articulus rostralis apicalis 454
– – – tracheal gill – artus branchialis apicalis 740
– labral bristle – seta labralis apicalis 818
– margin of clypeolabrum – clavus clypeolabralis apicalis 30
– – – clypeus – clavus clypealis apicalis 29
– – – coxa – clavus coxalis apicalis 60, 61
– – – frons – clavus frontalis apicalis 27

apical margin of frontoclypeus – clavus frontoclypealis apicalis **28**
– – – labrum – clavus labralis apicalis **31**
– – – pleurostoma – clavus pleurostomatalis apicalis **33**
– – – pronotum – clavus pronotalis apicalis 827
– – – trochanter – clavus trochanteralis apicalis **62, 63**
– part of mandible – pars apicalis mandibulae 920
– plate of tracheal gill – lamina branchialis apicalis 189
– pulvillus – pulvillus apicalis 363
– tibial bristle – seta tibialis apicalis 847
– trochanteral cavity – cavea trochanteralis apicalis **62, 63**
– ventral plate – lamina ventralis apicalis 503
apotorma – apotorma 566
arolium – arolium **66,** 263, 429–433, 698, 699, 772–775
articulation of mandible – articulatio mandibularis 1030, 1063
– – tracheal gill – articulatio branchialis 148
auditory organ – tympanum 247
– – of tibia – tympanum tibiale 243
auxilium – auxilium **66,** 394

barbula – barbula 614, 616
basal cervical sclerite – lamina cervicalis basalis 293
– condyle of mandible – condylus mandibularis basalis 388
– coxal cavity – cavea coxalis basalis **60, 61**
– denticle – denticulus basalis 88
– – of claw – denticulus unguis basalis 606
– femoral cavity – cavea femoralis basalis **64**
– joint – artus apicalis 797
– – of cercus – articulus cercalis basalis 364
– – – labial palp – artus palpi labialis basalis **37,** 771, 932
– – – maxillary palp – artus palpi maxillaris basalis **34,** 768, 931
– – – tracheal gill – artus branchialis basalis 740
– labral bristle – seta labralis basalis 818
– margin of clypeolabrum – clavus clypeolabralis basalis **30**
– – – clypeus – clavus clypealis basalis **29**
– – – coxa – clavus coxalis basalis **60, 61**
– – – frons – clavus frontalis basalis **27**
– – – frontoclypeus – clavus frontoclypealis basalis **28**
– – – labrum – clavus labralis basalis **31**
– – – pleurostoma – clavus pleurostomatalis basalis **33**
– – – pronotum – clavus pronotalis basalis 826
– – – trochanter – clavus trochanteralis basalis **62, 63**
– part of mandible – pars basalis mandibulae 920
– plate of tracheal gill – lamina branchialis basalis 189
– trochanteral cavity – cavea trochanteralis basalis **62, 63**
– ventral plate – lamina ventralis basalis 503
basicardo – basicardo 561, 563, 565

basiconic sensillum – sensillum basiconicum 566, 580, 584, 585, 915
basicostal suture – sutura basicostalis **59–61**
basicoxal cavity – cavea basicoxalis **60**
– furrow – fossa basicoxalis **60**
– suture – sutura basicoxalis **61**
basicoxite – basicoxa **47, 59–61**
basimentum – basimentum 806
basisternum – basisternum **46, 47, 58**
basistipes – basistipes 271–273, 317, 389, 390
basivalvula – basivalvula 249
branchial bristle – seta branchialis 189
– fold – plica branchialis 189
– hair – pilus branchialis 189
– spine – spina branchialis 189
– suture – sutura branchialis 189
– trachea – trachea branchialis 189
bristle – seta 907, 909, 910
– of abdominal leg – seta propodii 957, 958, 977
– – hypopodite – seta hypopodii 972
– – labial palp – seta palpi labialis 7, 372, 932
– – lateral lobe – seta lobi lateralis 171, 172
– – pygopod – seta pygopodii 852
– – scape – seta scapalis **41**
– – uropodite – seta uropodii 972
brush of hair – penicillus 570, 795
– – labial palp – scopa palpi labialis 5, 7
– – maxillary palp – scopa palpi maxillaris 4
buccula – buccula 454, 459, 482, 485, 486

campaniform sensillum – sensillum campaniformium 915, 916
campus – campus 614–617
capitate bristle – seta capitata 757
cardinal bristle – seta cardinalis 813, 1064
– cavity – cavea cardinalis **34**
cardinostipital suture – sutura cardinostipitalis **34**
cardo – cardo 4, 6, **22, 24, 34,** 77, 95, 111, 113, 136, 177, 198, 226, 227, 233–235, 252, 253, 255, 271–273, 287, 299, 301, 317–320, 336, 350, 352, 369, **514, 515,** 547, 560, 561, 563–565, 597, 677, 728, 729, 734, 745, 759, 761, 763, 765, 768, 813, 1020, 1023, 1024, 1027, 1064
caudal apex – apex caudalis 70
– bristle – macrochaeta caudalis 437
– bristle – seta caudalis 635
– comb – ctenidium caudale 858
– disc – discus caudalis 998
– filament – filamentum caudale 727
– hook – hamus caudalis 619

English—Latin

caudal leg – propodium caudale 785
– scolus – scolus caudalis 892
– spiracle – spiraculum caudale 1007, 1040, 1041
– sternite – sternum caudale 265
– tergite – tergum caudale 265
– tracheal gill – branchia caudalis 161
– vesicle – vesicula caudalis 750
central cavity – cavea centralis **528**
cephalic suture – sutura cephalica 1058, 1060
cephalotheca – cephalotheca 641, 647
cephalothorax – cephalothorax 479
cercal bristle – seta cercalis 221
– hair – pilus cercalis 221
– joint – articulus cercalis 221
– tracheal gill – branchia cercalis 161
cerculus – cerculus 184, 185
cercus – cercus **19, 69,** 90, 108, 109, 125, 126, 147, 149, 195, 196, 217–219, 224, 248–250, 266–268, 282, 285, 296, 313, 334, 346
cervical membrane – corium cervicale **25**
– region – regio cervicalis 306
– sclerite – lamina cervicalis 9, 398
cervix – cervix 270, 324, 326, **522**
chaetoparia – chaetoparia 566
chalaza – chalaza 906
cibarium – cibarium 1025, 1027
circumantennal suture – sutura circumantennalis 269, 271
clasp – tenaculum 395
clavate bristle – seta clavata 85, 797
claw – unguis 17, **57, 66,** 84–86, 106, 119, 120, 122, 142, 144, 215, 241–246, 263, 281, 308, 331, 333, 338, 339, 362, 363, 374–376, 394–397, 412–414, 428–433, 466, 467, 470, 472, 493–502, **525, 526,** 601, 602, 605, 673, 693, 696, 699, 700, 735, 748, 772–775, 779, 847–849, 949–953
clithrum – clithrum 566
clypeal bristle – seta clypealis 1019, 1059
– line – linea clypealis **29**
– membrane – corium clypeale 902
clypeofrontal suture – sutura clypeofrontalis 269
clypeogenal suture – sutura clypeogenalis **21, 27–30, 33,** 384, 763, 791
clypeolabral suture – sutura clypeolabralis **21, 28, 31, 33,** 286, **512,** 560, 670
clypeolabrum – clypeolabrum **30,** 272, 405, 419, 763, 791
clypeus – clypeus 2, **21, 22, 29,** 71–73, 110, 131, 135, 162, 197, 230, 231, 251, 270, 286, 288, 298, 300, 335, 347–349, 421, 454, 486, **513,** 560, 595, 648, 672, 673, 675, 690, 794, 901, 1019, 1021, 1059
comb of tracheal gill – ctenidium branchiale 148, 189
cone – conus 405, 406

coronal sinus of vertex − sinus verticalis coronalis 902
− suture − sutura coronalis **21, 23, 27, 28,** 131, 133, 165, 225, 269, 286, 298, 335, 347, **512,** 548, 595, 670, 676, 730, 744, 759, 760, 790, 791, 805, 806, 1019, 1022, 1059
corpotentorium − corpotentorium 676
corpus − corpus 89
corypha − corypha 566
coxa − coxa 17, **57–59,** 84, 106, 119–121, 144, 215, 241, 243, 245, 281, 308, 331, 332, 338, 374, 375, 394, 396, 427, 428, 466, 467, 498, **525,** 601–605, 693, 694, 697, 735, 748, 847–850, 949, 950
coxal articulation − articulatio coxalis **58, 59**
− carina − carina coxalis 308
− furrow − fossa coxalis **60**
− lobe − lobus coxalis **60, 61**
− membrane − corium coxale **58, 59, 525**
− suture − sutura coxalis **58, 60, 61,** 121, **526,** 951
− vesicle − vesicula coxalis 123
coxopleural process − processus coxopleuralis **58**
coxopleurite − coxopleuron **525,** 778
coxopodite − coxopodium 108, 738, 739, 742
crepis − crepis 566
crochet of abdominal leg − hamulus propodii 982
− − pseudopod − hamus pseudopodii 988
cylindrical joint of antenna − articulus antennalis cylindricus **40**

dens − dens 88, 89
denticle of claw − denticulus unguis 377
− − galeolacinia − denticulus galeolacinialis 562
− − pollex − denticulus pollicis 413
− − uncus − denticulus unci 854
− − denticulate area − area denticulata 1020
− spinneret − fusulus denticulatus 934
dexiotorma − dexiotorma 566
digitiform spine of tibia − spina tibialis digitiformis 245
discoidal spine − spina discoidalis 308, 309
dististipes − dististipes 273
dorsal abdominal ampulla − ampulla abdominalis dorsalis 627
− − bristle − seta abdominalis dorsalis 611, 799
− − gland − glandula abdominalis dorsalis 630, 968
− − papilla − papilla abdominalis dorsalis 799
− − protuberance − protuberantia abdominalis dorsalis 214
− − spine − spina abdominalis dorsalis 160, 186, 475, 885
− − tracheal gill − branchia abdominalis dorsalis 147, 870
− − tubercle − tuberculum abdominale dorsale 752
− appendage − appendicula dorsalis 184, 185
− area − area dorsalis 879, 880, 1040

dorsal bristle – seta dorsalis 983, 1035
– – of pygopod – seta pygopodii dorsalis 854
– brush – scopa dorsalis 999, 1044
– caudal papilla – papilla caudalis dorsalis 1041
– cervical sclerite – lamina cervicalis dorsalis **49**, 183, 337, 340, 347, 354, 355
– clypeus – clypeus dorsalis 384
– denticle – denticulus dorsalis 85
– femoral carina – carina femoralis dorsalis **64**
– gena – gena dorsalis 384, 387
– hook – hamus dorsalis 626
– intersegmental fold – plica intersegmentalis dorsalis 799
– line – linea dorsalis 880
– lobe of femur – lobus femoralis dorsalis **64**
– margin of femur – clavus femoralis dorsalis **64**
– papilla – papilla dorsalis 1007
– plate – lamina dorsalis 503
– scolus – scolus dorsalis 985
– setule – setula dorsalis 958
– spine – spina dorsalis 958, 985
– sternopleuron – sternopleuron dorsale **68**
– supracoxal sclerite – eupleuron **46**
– – – of prothorax – eupleuron prothoracale 98
– tergopleuron – tergopleuron dorsale **68**
– thoracic ampulla – ampulla thoracalis dorsalis 627
– tooth – dens dorsalis 920
– tubercle – tuberculum dorsale 1040
– valvula – valvula dorsalis 249
– verruca – verruca dorsalis 968
dorsocoxa – dorsocoxa **60, 61, 526**, 951
dorsolateral abdominal bristle – seta abdominalis dorsolateralis 611
– bristle – seta dorsolateralis 1035
– line – linea dorsolateralis 750
– margin of mesopleuron – clavus mesopleuralis dorsolateralis **44**
– – – metapleuron – clavus metapleuralis dorsolateralis **45**
– – – pronotum – clavus pronotalis dorsolateralis **43**
– – – propleuron – clavus propleuralis dorsolateralis **43**
dorsomedial abdominal bristle – seta abdominalis dorsomedialis 611
dorsopleural line – linea dorsopleuralis **527**

eighth abdominal epipleuron – epipleuron abdominale octavum 618
– – filament – filamentum abdominale octavum 738
– – hypopleuron – hypopleuron abdominale octavum 618
– – parasternite – parasternum abdominale octavum 1068
– – paratergite – paratergum abdominale octavum 1068
– – segment – segmentum abdominale octavum 618, 1068
– – spiracle – spiraculum abdominale octavum 629, 717

eighth abdominal sternite – sternum abdominale octavum 219, 283, 503, 784, 1068
– – tergite – tergum abdominale octavum 160, 503, 623, 741, 1068
– – tergopleuron – tergopleuron abdominale octavum 364
– epicranial bristle – seta epicranialis octava 595
– sternite – sternum octavum **69**
– tergite – tergum octavum **69**
empodial digitule – processus empodialis 85
empodium – empodium 85, 122, 696
epicranial bristle – seta epicranialis 792
– suture – sutura epicranialis 91, 251, 313, 320
epicranium – epicranium 72, 91–94, 251–253, 313, 320, 348–350, 353, 405, **512–515**, 547, 555, 557, 595, 596, 792, 793, 903, 904, 1023–1026
epifrons – epifrons **21, 22, 32, 47, 58, 59**
epimeron – epimeron **47, 58, 59**, 332, **525**
epipharynx – epipharynx 648
epipleural bristle – seta epipleuralis 954
epipleuron – epipleuron 954
epiproct – epiproctum 161, 382
episternum – episternum **58, 59**, 332, **525**
epistoma – epistoma 674, 675
epistomal bristle – seta epistomatalis 1019
– suture – sutura epistomatalis 271, 730, 792
– – of gula – sutura gularis epistomatalis **25**
epitorma – epitorma 566
epizygum – epizygum 566
euplantula – euplantula **66**, 246, 394
eusternum – eusternum **46**
eustipes – eustipes 317, 561
exodentate mandible – mandibula exodentata 1026
external → see also outer
– denticle of lacinia – denticulus laciniae externus 204
– lobe of glossa – lobus glossae externus 115
– – – paraglossa – lobus paraglossae externus 115, 299
– margin of tracheal gill – clavus branchialis externus 189
eye – oculus **19–24, 32**, 109, 110, 126, 130–135, 100–169, 175, 196, 197, 214, 224, 225, 229–232, 251, 267, 269, 270, 272, 273, 285–288, 297–300, 310, 313, 335, 336, 347–350, 353, 365–368, 384, 385, 405, 454, 458–460, 473, 481–486, 490, 546, 648, 649, 727, 791, 999, 1019, 1021, 1025

femoral bristle – seta femoralis 735, 950
– brush – scopa femoralis 144
– carina – carina femoralis 309
– denticle – denticulus femoralis 498
– furrow – fossa femoralis **64**, 215, 309
– lunule – lunula femoralis 242
– membrane – corium femorale 949

femur – femur 17, **57, 64,** 84, 106, 119, 120, 144, 215, 241–245, 281, 308, 309, 331, 338, 363, 374, 375, 394–396, 428, 466–469, 471, 498, 501, **525, 526,** 601–605, 693, 694, 697, 735, 748, 772, 847–850, 949–951
fifth abdominal spiracle – spiraculum abdominale quintum 789
– – sternite – sternum abdominale quintum 527
– – tergite – tergum abdominale quintum 750
– dorsal bristle – seta dorsalis quinta 954
– epicranial bristle – seta epicranialis quinta 595
– frontal bristle – seta frontalis quinta 595
– ocellus – ocellus quintus 912
– sternite – sternum quintum **69**
– tergite – tergum quintum **69**
filament of pygopod – filamentum pygopodii 738, 739, 741
first → see also anterior, fore
– abdominal epipleuron – epipleuron abdominale primum 596
– – filament – filamentum abdominale primum 736, 737
– – hypopleuron – hypopleuron abdominale primum 596
– – leg – propodium primum 655
– – segment – segmentum abdominale primum 18, 958
– – spiracle – spiraculum abdominale primum 596, 958
– – sternite – sternum abdominale primum 82, 211, 295, 296, 342, 380, 464, 465
– – tergite – tergum abdominale primum 82, 206, 216, 247, 278, 296, 322, 342, 359, **527,** 750, 1057
– adfrontal bristle – seta adfrontalis prima 902
– anal bristle – seta analis prima 972
– antennal joint – articulus antennalis primus 546, 583, 587
– anterior bristle – seta anterior prima 901, 911
– – frontoclypeal bristle – seta frontoclypealis anterior prima 805
– – vertical bristle – seta verticalis anterior prima 805
– clypeal bristle – seta clypealis prima 902
– dorsal bristle – seta dorsalis prima 954–958
– – vertical bristle – seta verticalis dorsalis prima 1059
– epicranial bristle – seta epicranialis prima 595
– frontal bristle – seta frontalis prima 595, 1019
– joint of rostrum – articulus rostralis primus 454
– labral bristle – seta labralis prima 1019
– lateral bristle – seta lateralis prima 901, 956
– – labral bristle – seta labralis lateralis prima 917
– – vertical bristle – seta verticalis lateralis prima 1060
– median labral bristle – seta labralis medialis prima 917
– mesothoracic fold – plica mesothoracalis prima 599
– – ring – anulus mesothoracalis primus 599, 944
– metathoracic ring – anulus metathoracalis primus 945
– ocellar bristle – seta ocellaris prima 911
– ocellus – ocellus primus 903, 904, 911–913
– posterior alveole – alveolus posterior primus 901, 904

first posterior bristle – seta posterior prima 901, 902, 904
– – – of pronotum – seta pronotalis posterior prima 826
– prothoracic ring – anulus prothoracalis primus 841, 943
– spiracular bristle – seta spiracularis prima 957, 958
– sternite – sternum primum **69**
– style – stylus primus 18
– tergite – tergum primum **69**
– thoracic ring – anulus thoracalis primus **525**
– ventral bristle – seta ventralis prima 956
– vertical bristle – seta verticalis prima 901, 1019
fixed coxa – coxa fixa **60**
flagellar joint – flagellomeron 275
flagellum – flagellum **40**, 386, 769
fold of abdominal ring – plica anuli abdominalis 799
forceps – forceps 267
fore → see also anterior, first
 – coxa – coxa anterior 3, 8–10, **42, 49, 50,** 78, 79, 97–99, 116, 145, 146, 163, 164, 175, 182, 183, 208, 232, 237, 239, 240, 257, 258, 260, 279, 280, 290, 293, 306, 310, 321, 324–326, 340, 343, 354, 356, 378, 381, 398, 401, 407, 435, 436, 462, 483, 484, **519, 522,** 598, 608, 609, 690, 776, 841, 844
 – coxal cavity – cavea coxalis anterioris **43**, 293, 307, **522,** 612, 781, 954
 – femur – femur anterius **42,** 141, 205, 290, 392, 461, 462, 598, 776
 – leg – pes anterior 1, **19, 20,** 70, 90, 109, 126, 160, 161, 195, 196, 224, 250, 267, 268, 285, 297, 313, 334, 346, 365, 383, 404, 408–411, 417, 439, 473, 479, 650
 – margin of mesonotum – clavus mesonotalis anterior **44,** 517
 – – – mesopleuron – clavus mesopleuralis anterior **44**
 – – – mesosternum – clavus mesosternalis anterior **44**
 – – – metanotum – clavus metanotalis anterior **45, 518**
 – – – metapleuron – clavus metapleuralis anterior **45**
 – – – metasternum – clavus metasternalis anterior **45**
 – – – pleurostoma – clavus pleurostomatalis anterior **33**
 – – – pronotum – clavus pronotalis anterior **43, 516**
 – – – propleuron – clavus propleuralis anterior **43**
 – – – prosternum – clavus prosternalis anterior **43**
 – posttrochanter – posttrochanter anterior 776
 – proleg – propes anterior **505–507,** 529, 635, 655, 672, 677, 690, 727, 736, 743, 752, 788, 789, 799, 879, 943, 946
 – subtrochanter – subtrochanter anterior 776
 – tarsus – tarsus anterior **42,** 141, 290, 598, 749, 776
 – tibia – tibia anterior **42,** 141, 290, 598, 776
 – trochanter – trochanter anterior **42,** 290, 356, 462
 – trochantin – trochantinus anterior **49,** 78, 98, 116, 293, 307, 321, 325, 326, 340, 343, 354
 – wing cover – theca alae anterioris **19, 20,** 126, 160, 161, 179, 180, 195, 196, 205, 207, 216, 224, 238, 297, 306, 310, 313, 323, 334, 361, 365, 417, 418, 439, 464, 473, 474, 490–492, **509,** 751, 756, 870

fourth abdominal laterosternite – laterosternum abdominale quartum **527**
– – poststernite – poststernum abdominale quartum 18
– – presternite – praesternum abdominale quartum 18
– – segment – segmentum abdominale quartum **505, 508**
– – sternite – sternum abdominale quartum 123
– – tergite – tergum abdominale quartum 1, 18, 285, 382, 529
– anal bristle – seta analis quarta 972
– dorsal bristle – seta dorsalis quarta 954–956
– epicranial bristle – seta epicranialis quarta 595
– frontal bristle – seta frontalis quarta 595
– interannular fold – plica interanularis quarta 701
– mesothoracic ring – anulus mesothoracalis quartus 599
– ocellus – ocellus quartus 912
– spiracular bristle – seta spiracularis quarta 957, 958
– sternite – sternum quartum **69**
– tergite – tergum quartum **69**
– ventral bristle – seta ventralis quarta 954, 956
free coxa – coxa libera **61**
frons – frons 2, **21, 22, 27,** 72, 73, 110, 135, 162, 166, 167, 169, 197, 225, 230–232, 251, 269, 270, 272, 286, 288, 298, 300, 313, 335, 347, 366, 367, 384, 419–421, 458, 485, **512, 513,** 546, 548, 549, 551, 560, 595, 648, 670, 673, 675, 690, 730, 744, 790–792, 794, 901–904, 1019, 1021–1024, 1057, 1059
frontal alveole – alveolus frontalis 901
– bristle – seta frontalis 2, 794, 901, 1059
– crest – crista frontalis 550
– groove – sulcus frontalis 648
– setule – setula frontalis 1059
– suture – sutura frontalis 72, 225, 251, 269, 271, 272, 298, 335, **512, 513,** 549, 560, 670, 690, 730, 744, 790, 792, 902, 1059
frontoclypeal suture – sutura frontoclypealis **21, 27, 29, 30,** 348, **512,** 560, 670, 672
frontoclypeus – frontoclypeus **28,** 91, 94, 133, 271, 460, 760, 805
frontogenal suture – sutura frontogenalis 384
frontovertical crest – crista frontoverticalis 546
– suture – sutura frontoverticalis **21,** 805, 1019, 1022
furca – furca 81
furcasternum – furcasternum **46, 47**

galea – galea 4, 6, **22, 24, 34, 35,** 95, 111, 177, 198, 202, 204, 226, 231, 233–235, 252, 253, 255, 270–273, 287–289, 299, 301, 317–320, 336, 352, 368, 369, 389, 390, **513,** 547, 561, 563, 670–672, 728, 729, 732, 734, 791, 930, 931, 1020
galea – lobus externus maxillae 514
galeal bristle – seta galealis 35
– cavity – cavea galealis 35
– comb – ctenidium galeale 35
– furrow – fossa galealis 35

galeal lobe – lobus galealis **35**
– plate – lamina galealis **35**
galeolacinia – galeolacinia 227, 229, 562, 793
gena – gena **21–23**, 73, 110, 132, 163, 164, 175, 229–231, 271, 286, 288, 298, 300, 335, 366, 367, 383, 458, 481–483, 485–487, 560, 648, 670, 674, 677, 728, 765, 791, 1023
genal alveole – alveolus genalis 904
– bristle – seta genalis 674, 1021
– line – linea genalis 32
– process – processus genalis 32
– region – regio genalis **32**, 71
– tuft – floccus genalis 131
genicular lobe – lobus genicularis 242
genital lobe – lobus genitalis 284
genu – genu **57, 65,** 241, 242, 331, 493, 748, 847
glandular pore – porus glandulae 438
– – of labium – porus glandulae labialis 371, 671, 672, 677
glomerate eyes – ocelli glomerati 799, 805
glossa – glossa 5, **22, 24, 37, 39,** 93, 114, 173, 199–201, 226–228, 236, 252, 256, 287, 289, 302, 314, 320, 336, 351, 368, 373, 391, **513**
glossal bristle – seta glossalis **39**, 373
– cavity – cavea glossalis 39
– comb – ctenidium glossale 39
– excision – rima glossae 173
– furrow – fossa glossalis 39
– lobe – lobus glossalis 39
– plate – lamina glossalis 39
gula – gula **23, 24,** 71, 72, 198, 226, 227, 252, 2566, 271, 273, 287, 350, 368, 454, 458, 459, **513–515,** 554–557, 671, 676, 728, 729, 745, 806, 813–815, 817, 1025, 1026
gulamentum – gulamentum 314, 320, 385
gular region – regio gularis **25**
– suture – sutura gularis 350, **514,** 728, 729, 745, 806, 816
gymnoparia – gymnoparia 566

halter – halter 649, 654
hamule – hamulus **528,** 786
haptolachus – haptolachus 566
haptomerum – haptomerum 566
head – caput 1, **19–24,** 70, 90, 109, 126, 160, 195, 214, 224, 250, 267, 285, 297, 313, 334, 346, 365, 383, 404, 418, 439, 473, **505–508, 512–515,** 529, 635–637, 642, 655, 727, 743, 752, 788, 799, 879, 880, 988, 999, 1001, 1027–1029, 1057
– cover – theca capitis **509, 510,** 654, 751, 870, 1069
– filament – filamentum capitis 885
– scolus – scolus capitis 892, 939
helus – helus 566

hind → see also posterior
- coxa – coxa posterior 15, 16, **42, 55, 56,** 82, 83, 105, 118, 183, 210, 216, 237–240, 259, 262, 264, 279, 280, 292, 295, 311, 329, 330, 342, 345, 360, 380, 381, 400, 403, 407, 435, 436, 464, 465, 492, **521, 524,** 600, 608, 609, 778, 843, 846
- coxal cavity – cavea coxalis posterioris **45,** 295, **524,** 612, 781, 956
- femur – femur posterius **42,** 143, 205, 259, 292, 324, 392, 600, 749, 778
- leg – pes posterior 1, **19, 20,** 70, 90, 109, 126, 160, 161, 195, 196, 224, 250, 267, 268, 285, 297, 313, 334, 346, 365, 383, 404, 408–411, 417, 439, 473, 650
- margin of mesonotum – clavus mesonotalis posterior **44, 517**
- – – mesopleuron – clavus mesopleuralis posterior **44**
- – – mesosternum – clavus mesosternalis posterior **44**
- – – metanotum – clavus metanotalis posterior **45, 518**
- – – metapleuron – clavus metapleuralis posterior **45**
- – – metasternum – clavus metasternalis posterior **45**
- – – pleurostoma – clavus pleurostomatalis posterior **33**
- – – pronotum – clavus pronotalis posterior **43, 516**
- – – propleuron – clavus propleuralis posterior **43**
- – – prosternum – clavus prosternalis posterior **43**
- posttrochanter – posttrochanter posterior 778
- proleg – propes posterior **505–507,** 529, 635, 655, 692, 735, 736, 743, 752, 788, 789, 799, 879, 892, 945, 948
- subtrochanter – subtrochanter posterior 778
- tarsus – tarsus posterior **42,** 143, 292, 600
- tibia – tibia posterior **42,** 143, 292, 600, 778
- trochanter – trochanter posterior **42,** 259, 292, 465, 492
- trochantin – trochantinus posterior **55,** 82, 104, 105, 118, 295, 329, 330, 342, 345, 360, 380, 612
- wing cover – theca alae posterioris **19, 20,** 126, 160, 161, 179, 180, 195, 196, 205, 207, 224, 264, 267, 278, 297, 313, 323, 334, 361, 365, 417, 418, 464, 473, 490, **509,** 649, 654, 870
hook – hamus 406
- of pseudopod – uncus pseudopodii 1036
- – pygopod – hamus pygopodii 739, 741
hypopharynx – hypopharynx **22,** 72, 73, 138, 198, 231, 546, 578, 648, 671, 672, 1024, 1025
hypopleural bristle – seta hypopleuralis 954
hypopleuron – hypopleuron 954
hypopodite – hypopodium **528,** 701, 972
hypostoma – hypostoma 674, 675, 729
hypostomal bridge – ponticulus hypostomatalis 793
- sclerite – scleritum hypostomatale 1027–1029
- suture – sutura hypostomatalis 555

incisor – dens incisus **26,** 74, 112, 176, 370, 684, 731, 747, 823, 1030, 1063
- lobe – dentes mediales 920
infra-anal plate – lamina infraanalis 184, 185

infraspiracular area – area infraspiracularis 701
inner → see also internal
– clypeal bristle – seta clypealis interior 1021
– frontal bristle – seta frontalis interior 1021
– occipital bristle – seta occipitalis interior 1021
– vertical brush of hair – penicillus verticalis interior 1022
intercervical sclerite – lamina intercervicalis 303
intermedial area – area intermedialis 1040
internal → see also inner
– denticle of lacinia – denticulus laciniae internus 204
– lobe of glossa – lobus glossae internus 115
– – – paraglossa – lobus paraglossae internus 115, 299
– margin of tracheal gill – clavus branchialis internus 189
intersegmental abdominal fold – plica abdominalis intersegmentalis 611, 701
– fold of metathorax – plica metathoracalis intersegmentalis 600
– – – prothorax – plica prothoracalis intersegmentalis 598
– – – thorax – plica thoracalis intersegmentalis **525**, 693
– membrane – membrana intersegmentalis 1, 437
– – of abdomen – corium abdominale intersegmentale 529
intersternite – intersternum 92
intestinal filament – filamentum intestinale 853

jugulum – iugulum 237
jugum – iugum 454, 458–460, 482, 485–487

katapleurite – katapleuron **46**

labial bristle – seta labialis 1062
– comb – ctenidium labiale 1026
– fissure – fissura labii 5
– lobe – lobus labialis 1023
– palp – palpus labialis 3, 5, 7, **22, 24,** 92–94, 110, 114, 115, 132, 139, 140, 195, 198–201, 225–228, 230–232, 236, 252, 253, 256, 270, 272, 273, 286–289, 299, 300, 302, 313, 314, 335, 336, 351, 353, 367, 368, 371, 372, 391, 406, 419, **513–515,** 546, 547, 559–561, 578, 597, 655, 670–673, 676, 677, 728, 729, 745, 758–761, 763–765, 793, 806, 813, 903, 918, 933, 1023, 1029, 1031, 1062
– plate – lamina labii 3
– sclerite – scleritum labiale 1027
– suture – sutura labialis **37**
labium – labium **22, 37,** 71–73, 76, 110, 161, 229–231, 268, 271, 288, 353, 367, 406, 419, 421, 459, 481–486, **513,** 546, 578, 690, 759, 765, 793, 903, 943, 946, 1020, 1026, 1060
labral alveole – alveolus labralis 917
– bristle – seta labralis 1059, 1065
– furrow – fossa labralis 818
– hair – pilus labralis 818

labral line – linea labralis **31**
– membrane – membrana labralis 805
– spine – spina labralis 818, 1065
labrum – labrum 2, **21, 22, 24, 31,** 71–73, 91, 92, 94, 110, 130–132, 135, 162, 175, 196–198, 225–227, 229–231, 251, 269–271, 273, 286, 288, 298, 300, 313, 335, 347, 349, 350, 353, 366–368, 385, 421, 454, 459, 483–487, **512–514,** 560, 567, 595, 597, 648, 655, 670–675, 690, 728, 730, 743, 744, 759, 790, 792, 794, 799, 805, 901–904, 943, 946, 1019, 1020, 1022–1027, 1057, 1059, 1060
lacinia – lacinia 4, 6, **22, 24, 34, 36,** 95, 111, 113, 177, 198, 202, 204, 226, 231, 233–235, 255, 271, 273, 289, 299, 301, 317, 319, 320, 336, 352, 368, 369, 389, 390, **513,** 561, 563, 672, 728, 734, 745, 768, 930, 931, 1020
lacinia – lobus internus 806, 813
lacinia – lobus internus maxillae **514**
lacinial bristle – seta lacinialis **36**
– cavity – cavea lacinialis **36**
– comb – ctenidium laciniale **36,** 204, 318, 352
– furrow – fossa lacinialis **36**
– lobe – lobus lacinialis **36**
– plate – lamina lacinialis **36**
laeotorma – laeotorma 566
lanceolate tarsal bristle – seta tarsalis lanceolata 952
last abdominal tergite – tergum abdominale ultimum 267
lateral abdominal papilla – papilla abdominalis lateralis 799
– – process – processus abdominalis lateralis 727
– – spine – spina abdominalis lateralis 160, 186
– – tracheal gill – branchia abdominalis lateralis 149, 804
– – tubercle – tuberculum abdominale laterale 752
– alveole – alvveolus lateralis 904
– area – area lateralis 1040
– bristle – setóa lateralis 902, 904, 911, 958, 1035
– – of mesonotum – seta mesonotalis lateralis 610, 828
– – – metanotum – seta metanotalis lateralis 829
– – – pronotum – seta pronotalis lateralis 610
– caudal papilla – papilla caudalis lateralis 1041
– cervical sclerite – lamina cervicalis lateralis 8, **49,** 257, 321, 340, 343, 350, 354, 356, 378, 677
– denticle – denticulus lateralis 85
– frontal suture – sutura frontalis lateralis **27**
– frontoclypeal bristle – seta frontoclypealis lateralis 805
– furcasternum of metathorax – furcasternum metathoracale laterale 16
– gular suture – sutura gularis lateralis **25**
– line – linea lateralis 800, 870, 879
– lobe – lobus lateralis 162–164, 166–168, 170, 173–175
– margin of clypeolabrum – clavus clypeolabralis lateralis **30**
– – – clypeus – clavus clypealis lateralis **29**
– – – frons – clavus frontalis lateralis **27**

lateral margin of frontoclypeus – clavus frontoclypealis lateralis 38
– – – labrum – clavus labralis lateralis 31
– – – mesonotum – clavus mesonotalis lateralis 44, 517
– – – mesosternum – clavus mesosternalis lateralis 44
– – – metanotum – clavus metanotalis lateralis 45, 518
– – – metasternum – clavus metasternalis lateralis 45
– – – pronotum – clavus pronotalis lateralis 43, 516
– – – prosternum – clavus prosternalis lateralis 43
– metanotal spine – spina metanotalis lateralis 392
– ocellus – ocellus lateralis 21, 130, 165, 197, 251, 298, 366, 791
– ostiole of wax gland – ostiolum cerae laterale 504
– plate of metanottum – lamina metanotalis lateralis 840
– pronotal plate – lamina pronotalis lateralis 840
– – spine – spina pronotalis lateralis 392
– scolus – scolus lateralis 985
– setule – setula lateralis 958
– tubercle – tuberculum laterale 1040
laterovertex – laterovertex 1022
leg – pes 57
ligula – ligula 92, 131, 132, 139, 140, 170, 299, 555, 557, 676, 728, 729, 745, 933
ligular excision – rima ligulae 170
lingua – lingua 92, 94, 115, 918
lobe – mala 136
– of pygopod – lobus pygopodii 853
longitudinal crest of pronotum – crista pronotalis longitudinalis 224
– furrow of clypeolabrum – fossa clypeolabralis longitudinalis 30
– – – frons – fossa frontalis longitudinalis 27
– – – frontoclypeus – fossa frontoclypealis longitudinalis 28
– line of clypeolabrum – linea clypeolabralis longitudinalis 30
– – – frons – linea frontalis longitudinalis 27
– – – frontoclypeus – linea frontoclypealis longitudinalis 28
– – – gula – linea gularis longitudinalis 25
– – – pronotum – linea pronotalis longitudinalis 355
lower epimeral sclerite – katepimeron 427
– – – of mesothorax – katepimeron mesothoracale 434, 435
– – – – metathorax – katepimeron metathoracale 434, 435
– – – – prothorax – katepimeron prothoracale 240, 279, 612
– episternal sclerite – katepisternum 47, 427
– – – of prothorax – katepisternum prothoracale 49, 279, 776
– margin of lateral lobe – clavus lobi lateralis inferior 174

mandible – mandibula 5, **22, 24, 26,** 71–73, 92, 94, 110, 131–135, 175, 196–198, 225–227, 229, 230–232, 251–253, 269–273, 286, 288, 298, 299, 313, 320, 335, 336, 347, 349, 350, 353, 366–368, 385, 389, 406, **505, 512, 513, 515,** 546, 547, 556, 557, 559, 560, 595, 597, 643, 648, 670, 672, 674,, 675, 677, 690, 728, 730,

744, 7445, 752, 757, 764, 790–794, 805, 806, 818, 901, 903, 943, 946, 988, 1019, 1022–1025, 1027–1029, 1031, 1034
mandibles – mandibulae 782
mandibular bristle – seta mandibularis **26,** 747, 792, 795, 823, 920, 1030, 1063
– canal – canalis mandibularis 576
– cavity – cavea mandibularis **26**
– condyle – condylus mandibularis 674, 731, 795, 824, 921
– denticle – denticulus mandibularis 388, 767
– furrow – fossa mandibularis **26,** 568, 767, 823
– lancet – spina mandibularis 419–421, 456, 765
– membrane – membrana mandibularis 388
– membrane – corium mandibulare 253, 560, 672, 676, 1019
– pore – porus mandibularis 576
– process – processus mandibularis 388
manubrium – manubrium 89
marginal bristle – seta marginalis 1068
maxilla – maxilla **34,** 71–73, 92, 94, 110, 132, 175, 229, 268, 288, **513,** 562, 648, 675, 690, 758–760, 762–764, 792, 918, 943, 946, 1019, 1026, 1058
maxillary apodeme – apodema maxillare 1064
– bristle – seta maxillaris 1060
– comb – ctenidium maxillare 1060
– lancet – spina maxillaris 420, 421, 456, 765
– lobe – lobus maxillaris 419, 420
– lobe – mala maxillaris 555, 565, 1024
– palp – palpus maxillaris 1–4, 6. **22, 24,** 72, 73, 77, 94, 95, 109–111, 113, 130–132, 136, 195, 197, 198, 202, 204, 225–227, 230, 232–234, 252, 253, 255, 267, 268, 270–273, 286–289, 299–301, 313, 317, 318, 320, 334–336, 347, 349, 350, 352, 353, 366–369, 389, 390, 419, **513–515,** 546, 547, 551, 552, 554, 558–564, 597, 655, 670–673, 676, 677, 728, 729, 732, 734, 745, 752, 791–794, 806, 813, 903, 930, 1020, 1021, 1023, 1024, 1031, 1058, 1060, 1064
maxillomandibular lancet – spina maxillomandibularis 761, 762
medial → see also median, middle
– lobe – lobus medialis 981
– plate of mesonotum – lamina mesonotalis medialis 840
– – – metanotum – lamina metanotalis medialis 840
median → see also medial, middle
– area – area medialis 1040
– bristle – seta medialis 1035
– – of mesonotum – seta mesonotalis medialis 610, 828
– – – pronotum – seta pronotalis medialis 610, 826
– cervical sclerite – lamina cervicalis medialis 8, 183, 677
– frontal bristle – seta frontalis medialis 1021
– labral bristle – seta labralis medialis 818
– laterocervical sclerite – lamina laterocervicalis medialis 303
– line of pronotum – linea pronotalis medialis 393
– ocellus – ocellus medialis **21,** 130, 165, 197, 251, 298, 366, 791

median ostiole of wax gland – ostiolum cerae mediale 504
- pronotal plate – lamina pronotalis medialis 840
- – spine – spina pronotalis medialis 392
- subjoint of tarsus – tarsomeron mediale **57, 66,** 362, 363, 494
- suture of mesonotum – sutura mesonotalis medialis 11
- – – metanotum – sutura metanotalis medialis 14
mediocranial groove – sulcus mediocranialis 648
membrane – membrana **67, 68**
- of labial plap – corium palpi labiale 236
- – maxillary palp – corium palpi maxillare 233
mental bristle – seta mentalis 371
- fissure – fissura mentalis 371
- membrane – membrana mentalis 761
- membrane – corium mentale 729
mentum – mentum **23,** 115, 200, 226–228, 236, 252, 253, 256, 270, 273, 299, 302, 371–373, 560, 561, 597, 648, 671, 677, 728, 729, 745, 806, 813, 903, 1023, 1024
meriston – meriston 276
meron – meron **47, 59**
mesofurca – mesofurca 344
mesonotal brush of hair – penicillus mesonotalis 828
- line – linea mesonotalis **51,** 278, 828
- suture – sutura mesosternalis 436
mesonotum – mesonotum 1, 11, 12, **42, 44, 51, 52,** 70, 80, 90, 100, 101, 109, 117, 161, 178, 183, 211, 250, 260, 261, 267, 278, 280, 285, 294, 322, 327, 337, 341, 346, 357, 381, 383, 392, 393, 399, 404, 411, 416, 434, 435, 463, **517, 520,** 529, 596, 608–610, 612, 727, 736, 737, 743, 749, 752, 780, 782, 799, 842
mesopleural suture – sutura mesopleuralis **52,** 294, 327, 341, 379, 464, **520**
- tracheal gill – branchia mesopleuralis 209
mesopleuron – mesopleuron 12, **42, 44,** 80, 81, 183, 259–261, 280, 327, 381, 463, 491, 609
mesosternal bristle – seta mesosternalis 399
- suture – sutura mesosternalis 436
mesosternum – mesosternum 12, **42, 44, 52,** 80, 101, 211, 239, 260, 261, 268, 279, 280, 324, 327, 328, 341, 379, 402, 436, **520,** 596, 597, 608, 609, 612, 947
mesostipes – mesostipes 273
mesothoracic acrotergite – acrotergum mesothoracale **51,** 97, 100, 101, 357, 434, 780
- anapleurite – anapleuron mesothoracale 12, 13, 101, **520**
- basicoxite – basicoxa mesothoracalis 259
- basisternum – basisternum mesothoracale 13, **53,** 81, 102, 117, 182, 209, 237, 240, 259, 294, 311, 344, 358, 399, 462–464, 491, **523,** 691, 781, 845, 944
- cover – theca mesothoracalis 870
- coxal tracheal gill – branchia coxalis mesothoracalis 211
- coxopleurite – coxopleuron mesothoracale **520, 523,** 842, 845, 944, 947

mesothoracic epimeron – epimeron mesothoracale **52,** 247, 294, 327, 328, 341, 379, 464, **520,** 612, 842
– epipleuron – epipleuron mesothoracale 596
– episternum – episternum mesothoracale **52,** 294, 327, 328, 341, 379, 434, 435, 464, **520,** 612
– furcasternum – furcasternum mesothoracale 13, **53,** 81, 102, 117, 182, 209, 237, 240, 259, 294, 311, 358, 463–465, 491, 691, 781
– hypopleuron – hypopleuron mesothoracale 596
– interpleurite – interpleuron mesothoracale 379, 381
– katapleurite – katapleuron mesothoracale 12, 101
– laterosternite – laterosternum mesothoracale 294
– laterotergite – laterotergum mesothoracale 11
– paranotum – paranotum mesothoracale 12, 13, **52**
– parasternite – parasternum mesothoracale 1058
– postnotum – postnotum mesothoracale 11, **51, 52,** 80, 117, 183, 260, 357, 780
– poststernite – poststernum mesothoracale 104, 105, 117, 209, 240, 294, 358, 399, **523,** 845, 944
– presternite – praesternum mesothoracale 12, 13, **53,** 81, 99, 102, 117, 209, 344, **523,** 691, 781, 944
– ring – anulus mesothoracalis 597
– scolus – scolus mesothoracalis 757, 789, 941
– segment – segmentum mesothoracale 637, 1001
– skin – coriaceum mesothoracale **517**
– spinasternum – spinasternum mesothoracale 311
– spiracle – spiraculum mesothoracale 11, **52,** 293
– sternellum – sternellum mesothoracale **53**
– sternopleuron – sternopleuron mesothoracale **53,** 182, 209, 240, 327, 328, 341, 358, 691
– sternum – sternum mesothoracale 1058
– tergite – tergum mesothoracale 1057
– tergopleuron – tergopleuron mesothoracale 379, 381, 608, 609
– tubercle – tuberculum mesothoracale 752
mesothorax – mesothorax **51–53,** 206, **505, 508, 517, 520, 523,** 538, 752, 780, 788, 879, 880, 1058
mesovertex – mesovertex 1022
metafurca – metafurca 345
metanotal brush of hair – penicillus metanotalis 829
– line – linea metanotalis 829
– spine – spina metanotalis 393
– suture – sutura metanotalis **54**
metanotum – metanotum 1, 14, 15, **42, 45, 54, 55,** 70, 82, 90, 103, 104, 109, 118, 178, 183, 211, 250, 260, 262, 280, 285, 295, 322, 329, 337, 342, 346, 359, 381, 383, 392, 393, 403, 404, 411, 416, 434, 435, 465, **518, 521,** 529, 596, 608, 609, 612, 727, 736, 737, 743, 749, 752, 780, 782, 799
metaparamere – metaparameron 284

metapleural suture – sutura metapleuralis **55**, 247, 295, 329, 342, 464, **521**
– tracheal gill – branchia metapleuralis 210
metapleuron – metapleuron 15, **42, 45,** 82, 83, 183, 259, 260, 262, 279, 280, 381, 465, 492, 608, 609
metasternal bristle – seta metasternalis 400, 403
metasternum – metasternum 15, **42, 45, 55,** 82, 104, 211, 239, 260, 262, 268, 279, 280, 311, 324, 329, 330, 342, 380, 403, 436, 465, 492, **521,** 596, 597, 608, 609, 612, 948
metatarsal fold – plica metatarsalis 362
– process – processus metatarsalis 376
– spine – spina metatarsalis 376
metatarsus – metatarsus **57, 66,** 120, 243, 246, 281, 333, 338, 362, 363, 374–376, 394–397, 493, 494, 496
metathoracic acrotergite – acrotergum metahoracale **54, 55,** 103, 104, 359
– anapleurite – anapleuron metathoracale 15, 104, **521**
– basicoxite – basicoxa metathoracalis 259
– basisternum – basisternum metathoracale 16, **56,** 83, 105, 118, 182, 210, 237, 240, 259, 295, 345, 360, 400, 464, **524,** 692, 781, 846, 945
– cover – theca metathoracalis 870
– coxal tracheal gill – branchia coxalis metathoracalis 211, 214
– coxopleurite – coxopleuron metathoracale **521, 524,** 843, 846, 945, 948
– epimeron – epimeron metathoracale **55,** 247, 295, 329, 330, 342, 380, 464, **521,** 612, 843
– epipleuron – epipleuron metathoracale 596, 843
– episternum – episternum metathoracale **55,** 247, 295, 329, 330, 342, 380, 434, 435, 464, **521,** 612, 842, 843
– furcasternum – furcasternum metathoracale **56,** 83, 105, 118, 182, 210, 237, 240, 259, 280, 295, 360, 464, 692
– hypopleuron – hypopleuron metathoracale 596, 843
– katapleurite – katapleuron metathoracale 15, 104
– laterosternite – laterosternum metathoracale 295
– paranotum – paranotum metathoracale 15, 16, **55**
– parasternite – parasternum metathoracale 1058
– postnotum – postnotum metathoracale 14, **54, 55,** 82, 118, 359, 434, 780
– poststernite – poststernum metathoracale 83, 105, 118, 210, 240, 360, 400, **524,** 781, 846, 945
– prepleurite – praepleuron metathoracale 608, 609
– presternite – praesternum metathoracale 15, 16, **56,** 83, 105, 118, 210, 345, **524,** 692, 945
– ring – anulus metathoracalis 597
– scolus – scolus metathoracalis 789, 941
– segment – segmentum metathoracale 637, 1001
– skin – coriaceum metathoracale **518,** 1058
– spinasternum – spinasternum metathoracale 16, 83
– spiracle – spiraculum metathoracale 14, **55,** 294
– sternellum – sternellum metathoracale **56**

metathoracic sternopleuron – sternopleuron metathoracale **56,** 118, 182, 210, 240, 329, 330, 342, 360, 403, 692
– sternum – sternum metathoracale 1058
– tergite – tergum metathoracale 1057
– tergopleuron – tergopleuron metathoracale 381
– tracheal gill – branchia metathoracalis 727
– tubercle – tuberculum metathoracale 752
metathorax – metathorax **54–56,** 206, **505, 508, 518, 521, 524,** 780, 788, 879, 880, 1058
metavertex – metavertex 298, 1022
microseta of mesonotum – microseta mesonotalis 828
– – metanotum – microseta metanotalis 829
middle → see also medial, median
– coxa – coxa medialis 12, 13, **42, 52, 53,** 80, 81, 100–102, 117, 164, 183, 209, 237–240, 247, 259, 261, 264, 279, 280, 291, 294, 310, 311, 327, 328, 341, 344, 358, 379, 381, 399, 402, 407, 435, 436, 463–465, 491, **520, 523,** 599, 608, 609, 777, 842, 845
– coxal cavity – cavea coxalis medialis **44,** 294, **523,** 612, 781, 955
– denticle of lacinia – denticulus laciniae medialis 235
– femur – femur mediale **42,** 142, 205, 259, 291, 392, 463, 599, 777
– furcasternum of metathorax – furcasternum metathoracale mediale 16
– joint – artus medialis 797
– – of labial palp – artus palpi labialis medialis **37,** 771
– – – maxillary palp – artus palpi maxillaris medialis **34,** 235
– – – tracheal gill – artus branchialis medialis 740
– leg – pes medialis 1, **19, 20,** 70, 90, 109, 126, 160, 161, 195, 196, 224, 250, 267, 268, 285, 297, 313, 334, 346, 365, 383, 404, 408–411, 417, 439, 473, 650
– posttrochanter – posttrochanter medialis 777
– proleg – propes medialis **505–507,** 529, 635, 655, 691, 727, 736, 743, 752, 788, 789, 799, 879, 892, 944, 947
– subtrochanter – subtrochanter medialis 777
– tarsus – tarsus medialis **42,** 142, 291, 599, 777
– tibia – tibia medialis **42,** 142, 291, 599, 749, 777
– trochanter – trochanter medialis **42,** 259, 291, 324, 463, 491, 599
– trochantin – trochantinus medialis **52,** 80, 101, 117, 294, 327, 328, 341, 344, 358, 379, 612
molar – mola 112, 370, 574, 685, 795, 825
– area – area molaris 75, 137
– crown – corona molaris **26,** 921
– lobe – lobus molaris 254, 687
mouth opening – porta epistomatalis 71
movable hook – denticulus mobilis 165, 170–174
mucro – mucro 88, 89
myocicatrix – myocicatrix 729

neck – collum 8,, 435, 672, 781
nesium – nesium 566
ninth abdominal segment – segmentum abdominale nonum 757, 799, 983, 1044
– – sternite – sternum abdominale nonum 248, 283, 716, 785
– – tergite – tergum abdominale nonum 219, 249, 266, 283, 437, 438, 738, 851, 853, 855
– sternite – sternum nonum **69**
– tergite – tergum nonum **69**
notum – notum **46, 47,** 427

occipital carina – carina occipitalis 197
– foramen – foramen occipitale **23–25,** 76, 226, 252, 287, 299, 303, 320, 336, 368, **513–515,** 552, 556, 557, 676, 728, 729, 759, 793, 806, 1026
– region – regio occipitalis **25**
– suture – sutura occipitalis 3, **22, 23, 25, 32, 33,** 71, 72, 226, 227, 252, 271, 273, 287, 299, 336, 554, 676, 728, 730, 744, 806
occiput – occiput 3, **22–25, 33,** 71, 76, 132, 197, 198, 226, 227, 252, 272, 273, 287, 299, 303, 320, 336, **513,** 546, 554, 556, 648, 671, 728–730, 744, 745, 758, 806, 1021
ocellar alveole – alveolus ocellaris 904
– area – area ocellaris 70–73, 76, 419–421, **513,** 635
– bristle – seta ocellaris 746
– protuberance – protuberantia ocellaris 757
ocellus – ocellus **512,** 548, 552, 655, 673, 690, 730, 746, 757, 758, 782
ocular area – area ocularis 790
– suture – sutura ocularis **32**
ocularium – ocularium 670
omma – omma 763
ommata – ommata 765
orbicular bristle – seta orbicularis 805
orbital margin of postocciput – clavus postoccipitalis orbitalis **25**
osmeterium – osmeterium 891
ostiole of wax gland – ostiolum cerae 480
outer → see also external
– clypeal bristle – seta clypealis exterior 1021
– frontal bristle – seta frontalis exterior 1021
– margin of lateral lobe – clavus lobi lateralis exterior 174
– occipital bristle – seta occipitalis exterior 1021
– vertical brush of hair – penicillus verticalis exterior 1022

palidium – palidium 614–616
palp – palpus 1028, 1029
palpifer – palpifer **34,** 226, 231, 234, 255, 271, 273, 368, 369, 547, 560, 597, 671, 728, 806, 813, 930
palpiform process – processus palpiformis 92, 94
palpiger – palpiger 5, 7, **37,** 228, 231, 256, 273, 289, 302, 676, 759, 763, 806, 933

palus – palus 614
papilla – papilla 905
papilliform process – processus papilliformis 1064
paraclypeus – paraclypeus 730, 744
paraglossa – paraglossa 5, 7, **22, 24, 37, 38,** 93, 114, 198–201, 226–228, 236, 251–253, 256, 272, 273, 287–289, 302, 314, 320, 336, 350, 351, 368, 371–373, 391, **513**
paraglossal bristle – seta paraglossalis 38
– cavity cavea paraglossalis 38
– comb – ctenidium paraglossale 38
– furrow – fossa paraglossalis 38
– lobe – lobus paraglossalis 38
– plate – lamina paraglossalis 38
paramentum – paramentum 199
paramere – parameron 284
paranotum – paranotum **46**
paraproct – paraproctum 161, 217, 219, 382
paraspiracular area – area paraspiracularis 701
parasternal bristle – seta parasternalis 1058
parastomal sclerite – scleritum parastomatale 1028
paratergal bristle – seta paratergalis 1057
paria – paria 566
paronychium – paronychium 493, 495, 500
paxilla – paxilla 851, 856
pedicel – pedicellus **40, 41,** 165, 277, 366, 386, 424, 454, 455, 459, 488, 489, 678, 680, 746, 769, 796, 805, 915, 916
pedicellar areola – areola pedicellaris 423
– bristle – seta pedicellaris 41
– membrane – corium pedicellare 41
pedium – pedium 566
penultimate abdominal sternite – sternum abdominale penultimum 268, 284
pharyngeal sclerite – scleritum pharyngeale 1028
pharynx – pharynx 1028
phoba – phoba 566
pinaculum – pinaculum 905
plantar surface – planta pulvillata **66,** 470
plate of tracheal gill – lamina branchialis 148
plegma – plegma 566
plegmatium – plegmatium 566
pleural membrane – corium pleurale **525**
– suture – sutura pleuralis **47, 58, 59,** 332, **525**
– tracheal gill – branchia pleuralis 212
pleurostoma – pleurostoma **22, 33,** 270, 670, 671, 674, 675, 690, 1059
pleurostomal bristle – seta pleurostomatalis 1059
– suture – sutura pleurostomatalis 270
pleuroventral line – linea pleuroventralis **527**

plumose bristle – seta plumosa 85, 797
poison canal – canalis venenatus 762
– gland – glandula venenata 762, 766
pollex – pollex 397, 413, 414
postantennal sclerite – lamina postantennalis 385
postclypeus – postclypeus 225, 269, 313, 366, 367, 481–485, 487, **512**, 670, 730, 744, 790, 792, 903, 1022
postcornu – postcornu 544, 634, 665, 719
posterior → see also hind
– bristle of mesonotum – seta mesonotalis posterior 610, 828
– – – pronotum – seta pronotalis posterior 610, 827
– clypeal bristle – seta clypealis posterior 1021
– condyle of mandible – condylus mandibularis posterior 33
– coronal suture – sutura coronalis posterior 793
– frontoclypeal bristle – seta frontoclypealis posterior 805
– lateral plate of mesonotum – lamina mesonotalis lateralis posterior 840
– laterocervical sclerite – lamina laterocervicalis posterior 303
– line of pronotum – linea pronotalis posterior 305
– lobe of pronotum – lobus pronotalis posterior 169
– – – trochanter – lobus trochanteralis posterior **62, 63**
– mediocervical sclerite – lamina mediocervicalis posterior 303
– mediolateral plate of mesonotum – lamina mesonotalis mediolateralis posterior 840
– – – – metanotum – lamina metanotalis mediolateralis posterior 840
– ocular margin – clavus ocularis posterior **21**
– spiracle – spiraculum posterius 1001
– sternal bristle – seta sternalis posterior 1058
– tentorial pit – fovea tentorialis posterior 92, 648, 729, 793
– tergal bristle – seta tergalis posterior 1057
– tergopleuron of metathorax – tergopleuron metathoracale posterius 608, 609
– thoracic notum – notum thoracale posterius 408
– valvula – valvula posterior 249
– vertical setule – setula verticalis posterior 1059
posterolateral bristle of pronotum – seta pronotalis posterolateralis 827
postfrons – postfrons **27, 32**
postfrontal suture – sutura postfrontalis **21, 22,** 91, 94, 110, 347, 348, 353
postgena – postgena **32,** 676, 806
postgenal suture – sutura postgenalis 676
postgula – postgula **25,** 76, 729
postlabium – postlabium 3, 5
postmental cavity – cavea postmentalis 37
postmentum – postmentum **24, 37,** 92–94, 114, 140, 164, 169, 170, 175, 198, 199, 287, 289, 350, 351, 368, 391, **514, 515,** 547, 558, 675, 761
postoccipital suture – sutura postoccipitalis **22, 23, 25,** 91, 94, 806
postocciput – postocciput **22–25, 33,** 91, 94, 287, 793, 806, 1021
postorbital suture – sutura postorbitalis 730

postnotum – postnotum **46, 47,** 427
postspiracular area – area postspiracularis 701, 717
– sclerite – lamina postspiracularis 341
– – of mesothorax – lamina postspiracularis mesothoracalis **52**
– – – metathorax – lamina postspiracularis metathoracalis **55**
posttrochanter – posttrochanter **63,** 84, 735, 847, 848, 850
– et subtrochanter – trochanter compositus **63**
preantennal brush of hair – penicillus praeantennalis 1022
preclypeus – praeclypeus 225, 269, 670
precoxal sclerite – scleritum praecoxale 307, 325, 326
prefrons – praefrons **27**
pregena – praegena **32**
pregula – praegula **25,** 76
prelabium – praelabium 3, 5, 7
prelabrum – praelabrum 791
premental bristle – seta praementalis 170
– cavity – cavea praementalis **37**
– fissure – fissura praementalis 201, 302, 314
– sclerite – scleritum praementi 675
prementum – praementum **24, 37,** 92–94, 114, 140, 162–164, 169, 170, 173–175, 198–201, 226, 227, 273, 287, 289, 302, 324, 320, 350, 351, 368, 385, 391, **514, 515,** 547, 558, 561, 597, 648, 675, 676, 728, 729, 745, 761, 768, 806, 813, 1024, 1027
presternal suture – sutura praesternalis **46, 47**
– – of mesothorax – sutura praesternalis mesothoracalis 13, **53,** 81
– – – metathorax – sutura praesternalis metathoracalis 16, **56,** 83
– – – prothorax – sutura praesternalis prothoracalis 10, **50**
presternite – praesternum **46, 47**
pretarsal bristle – seta praetarsalis 377, 775
pretarsus – praetarsus **57, 66,** 120, 246, 263, 281, 333, 338, 362, 363, 374, 376, 394–397, 493, 494, 496
prevertex – praevertex **32,** 298, 1022
profurca – profurca 343
proleg – propes **525**
pronotal crest – crista pronotalis **48**
– bristle – seta pronotalis 596
– brush of hair – penicillus pronotalis 827
– denticle – denticulus pronotalis 305
– furrow – fossa pronotalis 826
– line – linea pronotalis 205, 278, 827
– plate – lamina pronotalis 305
– spine – spina pronotalis 393
– suture – sutura pronotalis **48,** 78
pronotum – pronotum 1, 8, 9, **42, 43, 48, 49,** 70, 78, 90, 97, 98, 109, 116, 126, 160, 161, 163, 169, 178–181, 183, 195, 196, 205, 211, 214, 224, 230–232, 237–239, 250, 258, 267, 278–280, 285, 293, 297, 306, 310, 311, 313, 322, 323,

325, 334, 337, 340, 346–349, 354, 355, 381, 383, 392, 393, 398, 404, 411, 416, 434, 435, 439, 454, 460–462, 464, 473, 475, 483, 484, 490, 491, **506, 516, 519,** 529, 553, 608–610, 612, 672, 727, 736, 737, 743, 749, 752, 759, 780, 782, 799, 844

proplegmatium – proplegmatium 566

propleural suture – sutura propleuralis 49, 321, 325, 462, **519**

propleuron – propleuron 9, **42, 43,** 78, 183, 280, 325, 326, 381, 596–598, 608, 609, 672, 673

prosternal bristle – seta prosternalis 398

prosternum – prosternum 9, **42, 43, 49,** 78, 145, 175, 182, 211, 239, 258, 268, 280, 321, 324, 340, 354, 378, 401, 436, 461, 462, 464, **519,** 596, 597, 608, 609, 612, 672, 677, 946

prostheca – prostheca 137, 203, 570, 571, 575, 1030

prothoracic acrotergite – acrotergum prothoracale **48, 49,** 97, 98, 258, 780

– anapleurite – anapleuron prothoracale 8, 9, **519**

– basisternum – basisternum prothoracale 10, **50,** 79, 98, 99, 116, 208, 237, 240, 257, 293, 307, 343, 356, 398, **522,** 690, 781, 844, 943

– cover – theca prothoracalis 870

– coxal tracheal gill – branchia coxalis prothoracalis 208, 214

– coxopleurite – coxopleuron prothoracale **519,** 841, 943, 946

– epimeron – epimeron prothoracale **49,** 258, 293, 307, 321, 325, 340, 354, 378, 462, **519,** 776, 841

– episternum – episternum prothoracale 258, 293, 307, 321, 325, 340, 343, 354, 378, 435, 462, **519,** 612, 841

– fold – plica prothoracalis 598

– foramen – foramen prothoracale 307

– furcasternum – furcasternum prothoracale 10, **50,** 79, 98, 99, 116, 208, 237, 240, 257, 293, 307, 326, 690, 781

– hypopleuron – hypoleuron prothoracale 596

– katapleurite – katapleuron prothoracale 9

– laterosternite – laterosternum prothoracale 293

– membrane – corium prothoracale 546

– paranotum – paranotum prothoracale 237, 392

– postnotum – postnotum prothoracale **48, 49,** 78, 116

– poststernite – poststernum prothoracale 99, 101, 102, 116, 208, 293, 356, 398, **522,** 781, 844, 943

– presternite – praesternum prothoracale 10, **50,** 92, 93, 98, 99, 116, 208, 240, 280, 326, 343, 354, 356, **522,** 781, 943

– ring – anulus prothoracalis 597

– scolus – scolus prothoracalis 789

– segment – segmentum prothoracale 637, 1001

– skin – coriaceum prothoracale **516**

– spinasternum – spinasternum prothoracale 79, 237, 257

– spiracle – spiraculum prothoracale 655, 672, 673, 879, 954, 1007

– sternellum – sternellum prothoracale **50**

– sternopleuron – sternopleuron prothoracale **50,** 99, 116, 182, 208, 237, 240, 354, 690

prothoracic sternum – sternum protoracale 1058
– tergite – tergum prothoracale 1057, 1058
– tracheal gill – branchia prothoracalis 145, 146
– tubercle – tuberculum prothoracale 752
prothorax – prothorax **48–50, 505, 508, 516, 519, 522,** 538, 546, 673, 677, 760, 780, 788, 879, 880, 1058
pseudocercus – pseudocercus 283
pseudoculi – pseudoculi 1
pseudoculus – pseudoculus 2
pseudopod – pseudopodium 625
pternotorma – pternotorma 566
pulvillus – pulvillus 470, 494, 607
pulvinus – pulvinus 385
puparium – puparium **511,** 652–654
pygidium – pygidium 267, 282, 283
pygopod – pygopodium 738, 741, 799, 851–855

radicle – radicula **41**
ramus – ramus 89
rectal tracheal gill – branchia rectalis 994
respiratory siphon – tubus spiracularis 999, 1010, 1011, 1044, 1052
retinaculum – retinaculum 89, 552, 569, 795, 921
rhinarium – rhinarium 489
rostellum – rostellum 405
rostral furrow – fossa rostralis 459
rostrum – rostrum 1–3, 5, 456, 458, 460, 478
row of tibial pegs – series denticulorum tibialium 309

sagittal furrow of mesosternum – fossa mesosternalis sagittalis 13
salivary canal – canalis salivarius 933, 934
– gland – glandula salivaria 762
saw of apical denticle – serra denticuli apicalis 172
scapal membrane – corium scapale **41**
scape – scapus **40, 41,** 165, 197, 229–232, 253, 267, 269–272, 274, 298, 310, 366, 384, 386, 419–421, 423, 425, 454, 455, 457, 459, 481, 486, 488, 489, 560, 678–680, 728, 729, 746, 769, 770, 796, 805, 915
scent aperture – ostiolum odoriferum 439
second → see also middle
– abdominal laterotergite – laterotergum abdominale secundum **527**
– – paratergite – paratergum abdominale secundum 1057
– – sternite – sternum abdominale secundum 216
– – tergite – tergum abdominale secundum 123
– – tergopleuron – tergopleuron abdominale secundum 415
– adfrontal bristle – seta adfrontalis secunda 902
– anal bristle – seta analis secunda 972
– anterior bristle – seta anterior secunda 901, 911

second clypeal bristle – seta clypealis secunda 902
– dorsal bristle – seta dorsalis secunda 954–958
– – vertical bristle – seta verticalis dorsalis secunda 1059
– epicranial bristle – seta epicranialis secunda 595
– frontal bristle – seta frontalis secunda 595, 1019
– joint of rostrum – articulus rostralis secundus 454
– labral bristle – seta labralis secunda 1019
– lateral bristle – seta lateralis secunda 901, 956
– – labral bristle – seta labralis lateralis secunda 917
– – vertical bristle – seta verticalis lateralis secunda 1060
– median labral bristle – seta labralis medialis secunda 917
– thoracic fold – plica thoracalis secunda 599
– mesothoracic ring – anulus thoracalis secundus 599, 944
– metathoracic ring – anulus metathoracalis secundus 843, 945
– ocellar bristle – seta ocellaris secunda 911
– ocellus – ocellus secundus 912
– posterior alveole – alveolus posterior secundus 901, 904
– – bristle – seta posterior secunda 901, 902, 904
– – – of pronotum – seta pronotalis posterior secunda 826
– – vertical bristle – seta verticalis posterior secunda 805
– prothoracic fold – plica prothoracalis secunda 841
– – ring – anulus prothoracalis secundus 943
– spiracular bristle – seta spiracularis secunda 957
– sternite – sternum secundum **69**
– style – stylus secundus 18
– subjoint of tarsus – tarsomeron secundum 333
– tergite – tergum secundum **69**
– thoracic ring – anulus thoracalis secundus 693
– ventral bristle – seta ventralis secunda 954, 956
– vertical bristle – seta verticalis secunda 901, 1019
sensillum of labial palp – sensillum palpi labiale 226, 232, 236
– – maxillary palp – sensillum palpi maxillare 226, 231, 233–235
sensory appendage – appendicula sensillaris 583
– area – area sensillaris 581, 916
– bristle – sensillum chaeticum 489, 916
– cone – sensillum conicum 584
– hair – sensillum trichodeum 488, 580, 583, 915, 916, 930
– pit – caverna sensillaris 768
septula – septula 614, 616
seventh abdominal epipleuron – epipleuron abdominale septimum 618
– – filament – filamentum abdominale septimum 738, 741
– – hypopleuron – hypopleuron abdominale septimum 618
– – spiracle – spiraculum abdominale septimum **507**, 630, 655
– – sternite – sternum abdominale septimum 18, 285, 364, 503, 618
– – tracheal gill – branchia abdominalis septima 738
– epicranial bristle – seta epicranialis septima 595

seventh sternite – sternum septimum **69**
– tergite – tergum septimum **69**
simple bristle – seta simplex 85
– trochanter – trochanter simplex **62**
sinus of uncus – sinus unci 855
siphuncle – siphunculus 476
sixth abdominal segment – segmentum abdominale sextum 161, 383
– – spiracle – spiraculum abdominale sextum 879, 898
– – sternite – sternum abdominale sextum 108
– epicranial bristle – seta epicranialis sexta 595
– sternite – sternum sextum **69**
– tergite – tergum sextum **69**
sole – planta 742, 786, 798, 982
spathulate bristle – seta spathulata 437
spatula – spatula 982
spatulate bristle – seta spatulata 488, 502, 772
spinasternum – spinasternum **46**
spine – spina 908
– of claw – spina unguis 377
– – pollex – spina pollicis 414
– – stylus – spina styli 124
spinneret – fusulus 813, 903, 918, 933
spiracle – spiraculum 79, 98, 101, 104, 125, 258, 283, 306, 325, 327, 329, 381, 465, 483, 598, 610, 618, 625, 627, 707, 782
spiracular area – area spiracularis 701, 717
– sclerite – lamina spiracularis 608, 609
– spur – calcar spiraculare 629
stemmata – stemmata 744, 745, 792
sternacostal suture – suture sternacostalis **47**
– – of mesothorax – sutura sternacostalis mesothoracalis **53**
– – – metathorax – sutura sternacostalis metathoracalis 16, **56**
– – – prothorax – sutura sternacostalis protoracalis **50**, 79
sternal tracheal gill – branchia sternalis 213
sternellum – sternellum **47**
sternite – sternum **67, 68,** 427
sternopleural suture of prothorax – sutura sternopleuralis prothoracalis **50**
sternopleuron – sternopleuron **47, 67,** 121
stigma – stigma 640
stipes – stipes 4, 6, **22, 24, 34,** 77, 95, 111, 113, 136, 177, 198, 202, 204, 226, 227, 233–235, 252, 253, 255, 272, 286, 289, 299, 301, 318–320, 336, 350, 352, 368, 369, 389, 390, **514, 515,** 547, 560, 561, 563–565, 597, 677, 728–730, 732, 734, 745, 759, 761, 763, 765, 768, 806, 813, 903, 931, 1020, 1023, 1024, 1064
stipital bristle – seta stipitalis 813, 1064
– cavity – cavea stipitalis **34**
– membrane – corium stipitale 561
– process – processus stipitalis 233

style – stylus 107, 108, 123, 125, 248, 249, 477
styloconic sensillum – sensillum styloconicum 915, 919
subanal appendage – appendicula subanalis 717
– lobe – lobus subanalis 248, 266, 282
– plate – lamina subanalis 219, 266, 296, 382, 614, 617, 715, 717
subantennal sclerite – lamina subantennalis 385
– suture – sutura subantennalis **21, 22, 32,** 229, 335
subapical denticle – denticulus subapicalis 88, 170, 172, 173
– – of claw – denticulus unguis subapicalis 606
– – – lacinia – denticulus laciniae subapicalis 234, 235, 318
– pulvillus – pulvillus subapicalis 363
– tooth – dens subapicalis **26,** 112
subcostal suture – sutura subcostalis **61**
– tracheole – tracheola subcostalis 189
subcoxa – subcoxa **61,** 693, 697
subcoxal suture – sutura subcoxalis **61**
– tracheal gill – branchia subcoxalis 144
subdorsal area – area subdorsalis 879, 880, 1040
– line – linea subdorsalis 879, 880
subgenal suture – sutura subgenalis **32, 33**
subhypostomal sclerite – scleritum subhypostomatale 1028
submarginal suture of mesothorax – sutura submarginalis mesothoracalis **523**
– – – metathorax – sutura submarginalis metathoracalis **524**
– – – prothorax – sutura submarginalis prothoracalis **522**
submental bristle – seta submentalis 813
– fissure – fissura submentalis 813
submentum – submentum **23,** 200, 226, 227, 299, 302, 560, 561, 597, 648, 676, 728, 745, 806, 813, 903, 1023
subocellar alveole – alveolus subocellaris 904
– suture – sutura subocellaris 71, 72
subocular suture – sutura subocularis **22, 32,** 288, 335
subspiracular area – area subspiracularis 701, 702, 704, 711, 717, 879
– sclerite – lamina subspiracularis 258, 609, 709
– – of mesothorax – lamina subspiracularis mesothoracalis **52**
– – – metathorax – lamina subspiracularis metathoracalis **55**
– scolus – scolus subspiracularis 969
– spine – spina subspiracularis 958
– tracheal gill – branchia subspiracularis 971
– verruca – verruca subspiracularis 968
subtrochanter – subtrochanter **63,** 84, 735, 847, 848, 850
subventral line – linea subventralis 879
superior tentorial pit – fovea tentorialis superior 91, 648, 672
superlingua – superlingua 138
supra-anal lobe – lobus supraanalis 248
– plate – lamina supraanalis 184, 217–219, 621, 715, 717
suprapedal area – area suprapedis **528,** 701, 702, 717

suprapedal sclerite – lamina suprapedis 709
supraspiracular area – area supraspiracularis 701, 879, 880
– scolus – scolus supraspiracularis 970
– spine – spina supraspiracularis 958
– tracheal gill – branchia supraspiracularis 971
– verruca – verruca supraspiracularis 968
syphon – sipho 422, 453

tail – cauda 90, 415, 437, 438, 504, **505, 506,** 784, 884
tarsal apophysis – apophysis tarsalis 414
– bristle – seta tarsalis 339, 412, 429, 495, 500, 502, 774, 950
– membrane – corium tarsale **526,** 950
– pulvillus – pulvillus tarsalis 263
– sclerite – scleritum tarsale 414
tarsungulus – tarsungulus 603, 604
tarsus – tarsus 17, **57, 66,** 106, 119, 120, 122, 144, 215, 243–245, 308, 331, 412–414, 428, 466, 467, 471, 472, 495, 497–499, 501, **525, 526,** 601, 605, 693, 694, 697, 735, 748, 774, 847, 848, 850, 950–953
teges – teges 616, 617
tegillum – tegillum 614
tegminal cover – theca tegminis 264, 267, 279
tegula – tegula 361
telson – telson 18, 195, 538
temple – tempus **22–24, 32,** 132, 229, 231, 232, 253, 287, 288, 299, 300, 367, 368, 670, 671, 676
tenth abdominal segment – segmentum abdominale decimum 655, 738, 1066
– – sternite – sternum abdominale decimum 125, 248, 282, 715
– – tergite – tergum abdominale decimum 125, 185, 219, 248, 282, 296, 783, 853
– sternite – sternum decimum **69**
– tergite – tergum decimum **69**
tentorial pit – fovea tentorialis 286
tentorium – tentorium 421
tergal bristle – seta tergalis 851, 853
tergite – tergum **67, 68**
tergopleuron – tergopleuron 67
terminal antennal bristle – seta antennalis terminalis 1061
– bristle – seta terminalis 799
– filament – filum terminale 109, 125, 126, 147, 150, 858
– tracheole – tracheola terminalis 189
thin hair – cilium 906
third → see also hind
– abdominal parasternite – parasternum abdominale tertium 1058
– – pleurite – pleuron abdominale tertium **527**
– – ring – anulus abdominalis tertius 701
– – scolus – scolus abdominalis tertius 789

third abdominal segment – segmentum abdominale tertium 799, 1001
– – spiracle – spiraculum abdominale tertium **506**, 750
– – sternite – sternum abdominale tertium 123, 324, 382, 783, 1058
– – tergite – tergum abdominale tertium 89, 108, 415
– anal bristle – seta analis tertia 972
– anterior bristle – seta anterior tertia 904, 911
– dorsal bristle – seta dorsalis tertia 954–956
– – vertical bristle – seta verticalis dorsalis tertia 1059
– epicranial bristle – seta epicranialis tertia 595
– frontal bristle – seta frontalis tertia 595
– joint of rostrum – articulus rostralis tertius 454
– labral bristle – seta labralis tertia 1019
– lateral bristle – seta lateralis tertia 956
– – labral bristle – seta labralis lateralis tertia 917
– median labral bristle – seta labralis medialis tertia 917
– mesothoracic ring – anulus mesothoracalis tertius 599, 944
– metathoracic fold – plica metathoracalis tertia 600
– ocellar bristle – seta ocellaris tertia 911, 913
– ocellus – ocellus tertius 912, 914
– spiracular bristle – seta spiracularis tertia 957
– sternite – sternum tertium **69**
– style – stylus tertius 18
– subjoint of tarsus – tarsomeron tertium 333
– tergite – tergum tertium **69**
– ventral bristle – seta ventralis tertia 956, 957
– vertical bristle – seta verticalis tertia 901
thoracic brush of hair – penicillus thoracalis 999
– cover – theca thoracalis **509**, 654, 1069
– foramen – foramen thoracale 609, 781
– notum – notum thoracale 410
– process – processus thoracalis 1000
– pseudopod – pseudopodium thoracale 988, 1011
– scolus – scolus thoracalis 888, 892
– segment – segmentum thoracale **46, 47, 525**
– skin – coriaceum thoracale 799
– spiracle – spiraculum thoracale 247, 529, 596, 693
– sternum – sternum thoracale 407
thorax – thorax 1, **19, 20, 42**, 70, 90, 109, 126, 160, 195, 224, 250, 267, 285, 297, 313, 334, 346, 365, 383, 404, 418, 439, 473, **505–508**, 529, 635, 636, 642, 655, 727, 743, 752, 788, 880, 999, 1057
tibia – tibia 17, **57, 65**, 106, 119, 120, 144, 215, 241, 243–245, 281, 308, 309, 331, 338, 363, 374–376, 394–397, 412–414, 428, 466–468, 471, 472, 493, 496–499, 501, **525, 526**, 601, 603–605, 693, 694, 697, 735, 748, 847, 848, 850, 950, 951
tibial bristle – seta tibialis **65**, 376, 412, 735, 950
– brush – scopa tibialis 144, 748
– denticle – denticulus tibialis **65**

tibial furrow – fossa tibialis **65**
– hair – cilium tibiale **65**
– membrane – corium tibiale **526**, 950
– process – processus tibialis 499
– spine – spina tibialis **65**, 241, 362, 472, 493
– spur – calcar tibiale **57, 65, 66,** 308, 309, 376
tibiotarsal bristle – seta tibiotarsalis 949
– spine – spina tibiotarsalis 242
tibiotarsus – tibiotarsus 84, 85, 242, 469, 602, 695, 772, 773, 949
trachea – trachea 148
tracheal gill – branchia 742
– trunk – truncus trachealis 189
transversal frontal suture – sutura frontalis transversalis **27, 28**
– furrow of clypeolabrum – fossa clypeolabralis transversalis **30**
– – – clypeus – fossa clypealis transversalis 670
– – – frons – fossa frontalis transversalis **27**
– – – frontoclypeus – fossa frontoclypealis transversalis **28**
– line of clypeolabrum – linea clypeolabralis transversalis **30**
– – – frons – linea frontalis transversalis **27**
– – – frontoclypeus – linea frontoclypealis transversalis **28**
– – – gula – linea gularis transversalis **25**
– – – pronotum – linea pronotalis transversalis 355
– suture – sutura transversalis 299
– – of pronotum – sutura pronotalis transversalis 305, 306
– – – vertex – sutura verticalis transversalis **21**
trochanter – trochanter 17, **57,** 106, 119, 120, 144, 215, 241, 243, 245, 281, 308, 309, 331, 338, 374, 375, 394, 396, 428, 466, 467, 498, **525, 526,** 601–605, 693, 697, 748
trochanteral suture – sutura trochanteralis **63**
trochantin – trochantinus **46, 58, 59,** 121, 332
tuft – floccus 998

ungual carina – carina unguis 263
– lobe – lobus unguis 377
– membrane – corium unguis 377
unguitractor plate – lamina unguitractoralis 122, 432
upper margin of lateral lobe – clavus lobi lateralis superior 174
urogomphus – urogomphus 621–625
uropodite – uropodium **528**, 701, 972

valve – valva 503
vasiform orifice – orificium vasiforme 480
ventral abdominal ampulla – ampulla abdominalis ventralis 789
– – bristle – seta abdominalis ventralis 799
– – tracheal gill – branchia abdominalis ventralis 804, 870
– ampulla – ampulla ventralis 713

ventral appendage – appendicula ventralis 184, 185
- area – area ventralis 879, 1040
- bristle – seta ventralis 977, 1035
- – of pygopod – seta pygopodii ventralis 854
- brush – scopa ventralis 999, 1044
- caudal papilla – papilla caudalis ventralis 1041
- – cervical sclerite – lamina cervicalis ventralis **49**, 175, 183, 257, 321, 340, 350, 354, 356
- clypeus – clypeus ventralis 385
- denticle – denticulus ventralis 85
- femoral carina – carina femoralis ventralis **64**
- gena – gena ventralis 385
- intersegmental fold – plica intersegmentalis ventralis 799
- line – linea ventralis 76, 79, 81, 83, 87
- lobe of femur – lobus femoralis ventralis **64**
- margin of femur – clavus femoralis ventralis **64**
- papilla – papilla ventralis 1007
- setule – setula ventralis 958
- spine – spina ventralis 958
- sternopleuron – sternopleuron ventrale **68**
- tergopleuron – tergopleuron ventrale **68**
- tooth – dens ventralis 920
- tube – tubus ventralis 82
- tubercle – tuberculum ventrale 1040
- verruca – verruca ventralis 968
- vertical bristle – seta verticalis ventralis 1060

ventrocoxa – ventrocoxa **60, 61, 526,** 951
ventrolateral bristle – seta ventrolateralis 1035
- margin of mesopleuron – clavus mesopleuralis ventrolateralis **44**
- – – metapleuron – clavus metapleuralis ventrolateralis **45**
- – – propleuron – clavus propleuralis ventrolateralis **43**

vertex – vertex 1–4, **21–23,** 71–73, 110, 132, 135, 162, 163, 165, 167–169, 197, 225–227, 229–231, 269–273, 286, 288, 299, 300, 335, 336, 347, 366–368, 419, 420, 458, 460, 481–483, 485, 486, **513,** 546, 550, 560, 648, 670–673, 675, 676, 690, 730, 744, 745, 758–761, 765, 790, 791, 793, 794, 805, 806, 901, 1019, 1021, 1059, 1060

vertical bristle – seta verticalis 2, 794
- crest – crista verticalis 546
- furrow – fossa verticalis **21,** 670, 673, 676, 790
- groove – sulcus verticalis 648
- suture – sutura verticalis 251
- tuft – corypha verticalis 805

vesicle – vesicula 87, **528,** 739, 742, 786, 797, 798, 972

wing cover – theca alae **510**

zygum – zygum 566

SELECTED REFERENCES

Aspöck, H. U. (1971): Raphidioptera. — *Handb. Zool.*, **4** (2), 15:1–48.
Bartkowska, K. (1965): Morfologia larwy pchly Typhloceras poppei Wagn. (Aphaniptera, Hystrichopsyllidae). — *Annls zool., Warsz.*, **23** (9): 237–250.
Bartkowska, K. (1972): Morfologia larwy Rhadinopsylla (Actenophthalmus) integella Jord. et Roths. (Siphonaptera). — *Polskie Pismo ent.*, **42**: 535–543.
Beck, H. (1960): Die Larvalsystematik der Eulen (Lepidoptera). — *Abh. Larvalsyst. Insekten*, **4**: 1–406.
Beier, M. (1937): Suctoria (Siphonaptera, Aphaniptera). — *Handb. Zool.*, **4** (2), 9: 1999–2039.
Beier, M. (1968): Phasmida. — *Handb. Zool.*, **4** (2), 6: 1–56.
Beier, M. (1968): Mantodea. — *Handb. Zool.*, **4** (2), 4: 1–47.
Beier, M. (1969): 5. Klassifikation. — *Handb. Zool.*, **4** (2), 9: 1–17.
Beier, M. (1972): Saltatoria (Grillen und Heuschrecken). — *Handb. Zool.*, **4** (2), 17: 1–217.
Beier, M. (1974): Blattaria. — *Handb. Zool.*, **4** (2), 22: 1–127.
Bertrand, H. (1954): *Les insectes aquatiques d'Europe*. XXX–XXXI. — Paul Lechevalier Éditeur, Paris, I: 556 pp.; II: 547 pp.
Bretfeld, G. (1963): Zur Anatomie und Embryologie der Rumpfmuskulatur und der Abdominalen Anhänge der Collembolen. — *Zool. Jb., Anatomie*, **80**: 309–384.
Bruckmoser, P. (1965): Embryologische Untersuchungen über den Kopfbau der Collembole Orchesella villosa L. — *Zool. Jb., Anatomie*, **82**: 299–364.
Buckup, L. (1958): Der Kopf von Myrsidea cornicis (de Geer) (Mallophaga). — *Zool. Jb., Anatomie*, **77**: 241–288.
Butler, E. A. (1923): *A biology of the British Hemiptera-Heteroptera*. — H. F. & G. Witherby, London, 682 pp.
Crampton, G. C. (1909): Contribution to the comparative morphology of the thoracic sclerites of insects. — *Proc. Acad. nat. Sci. Philad.*, **61**: 3–54.
Elbel, R. E. (1951): Comparative studies on the larvae of certain species of fleas (Siphonaptera). — *J. Parasit.*, **37**: 119–128.
Fischer, M. (1969): Die Verwandlung der Insekten. — *Handb. Zool.*, **4** (2), 8: 1–68.
Francois, J. (1970): Squellete et musculature céphalique de Campodea chardardi Condé (Diplura). — *Zool. Jb., Anatomie*, **87**: 331–376.
Geisthardt, M. (1979): Skelet und Muskulatur des Thorax der Larven und Imagines von Lamprohiza splendidula (L.) unter Berücksichtigung der Larve und der weiblichen Imago von Lampyris noctiluca (L.) (Coleoptera). — *Zool. Jb., Anatomie*, **101**: 472–536.
Ghilarov, M. S. (ed.) — Гиляров, М. С. (редактор) (1964): *Определитель обитающих в почве личинок насекомых*. — АН изд. "Наука", Москва, 919 пп.
Giles, E. T. (1963): The comparative external morphology and affinities of the Dermaptera. — *Trans. R. ent. Soc. Lond.*, **115**: 95—164.
Grassé, P. P. (1949–1951): *Traité de Zoologie. Anatomie, Systematique, Biologie*. — Masson et Cie Éditeurs, **IX**: 1117 pp. **X**: 1940 pp.
Günther, K. und Herter, K. (1974): Dermaptera. — *Handb. Zool.*, **4** (2), 23: 1–158.
Hayashi, N. (1980): Illustrations for identification of larvae of the Cucujoidea (Coleoptera). — *Mem. Educ. Inst. Priv. Schools Japan*, **72**: 1–53.
Hendel, F. (1936): Diptera. — *Handb. Zool.*, **4** (2), 9: 1757–1998.
Hennig, W. (1968): *Die Larvenformen der Dipteren*. — Akademie-Verlag, Berlin, 458 pp.
Illies, J. (1968): Ephemeroptera. — *Handb. Zool.*, **4** (2), 7: 1–63.
Imms, A. D. (1957): *A general textbook of entomology*. — Methuen & Co. Ltd., London, 886 pp.
Janetschek, H. (1970): Protura. — *Handb. Zool.*, **4** (2), 14: 1–72.

Kaltenbach, A. (1968): Embiodea. — *Handb. Zool.,* **4** (2), 1: 1-29.
Kéler, S. von (1969): Ordnung Mallophaga (Federlinge und Haarlinge). — *Handb. Zool.,* **4** (2): 1-72.
Kinzelbach, R. (1966): Zur Kopfmorphologie der Fächerflügler (Strepsiptera). — *Zool. Jb., Anatomie,* **84**: 559-684.
Kinzelbach, R. (1971): Strepsiptera. — *Handb. Zool.,* **4** (2), 15: 1-68.
Kinzelbach, R. és Kaszab, Z. (1977): Legyezőszárnyúak — Strepsiptera. — In: *Magyarország Állatvilága (Fauna Hungariae),* **X** (10). Akadémiai Kiadó, Budapest, 54 pp.
Klapalek, F. (1909): Plecoptera. — In: *Die Süsswasserfauna Deutschlands,* **8**: 33—95.
Krauss, H. A. (1911): *Monographie der Embien.* — Schweizerbart'sche Verlagsbuchhandlung, Stuttgart, 78 pp.
Krivnosheina, N. P.— Кривошеина, Н. П. (1969): *Онтогенез и эволюция двукрылых насекомых.* — АН изд. "Наука", Москва, 291 пп.
Lyon, H. (1915): Notes on the cat flea (Ctenocephalus felis Bouché, Siphonaptera). — *Psyche,* **22**: 124-132.
MacSwain, J. W. (1956): *A classification of the first instar larvae of the Meloidae (Coleoptera).* — University of California Press, Berkeley and Los Angeles, 182 pp.
Malicky, H. (1973): Trichoptera. — *Handb. Zool.,* **4** (2), 21: 1-114.
Matsuda, R. (1965): Morphology and evolution of the insect head. — *Mem. Am. entomol. Inst.,* **4**: 1-334.
Matsuda, R. (1970): Morphology and evolution of the insect thorax. — *Mem. entomol. Soc. Can.,* **76**: 1-431.
Matsuda, R. (1976): *Morphology and evolution of the insect abdomen.* — Pergamon Press, New York, 534 pp.
Mayer, C. (1954): Vergleichende Untersuchungen am Skelett-Muskelsystem des Thorax der Mallophagen unter Berücksichtigung des Nervensystems. — *Zool. Jb., Anatomie,* **74**: 77-131.
Meixner, J. (1934): Coleopteroidea. — *Handb. Zool.,* **4** (2), 2: 1037-1340.
Mickoleit, G. (1961): Zur Thoraxmorphologie der Thysanoptera. — *Zool. Jb., Anatomie,* **79**: 1-92.
Neuffer, G. (1954): Die Mallophagenhaut und ihre Differenzierungen. — *Zool. Jb., Anatomie,* **73**: 450-518.
Parker, H. L. (1924): Recherches sur les formes post-embryonnaires des Chalcidiens (Hymenoptera). — *Annls. Soc. ent. Fr.,* **93**: 261-379.
Peterson, A. (1953): *Larvae of insects. Part II. Coleoptera, Diptera, Neuroptera, Siphonaptera, Mecoptera, Trichoptera.* —. Edwards Brothers, Inc., Ann Arbor, Michigan, 416 pp.
Peterson, A. (1959): *Larvae of insects. Part I. Lepidoptera and plant infesting Hymenoptera.* Edwards Brothers. Inc., Ann Arbor, Michigan, 315 pp.
Plachter, H. (1979): Zur Kenntnis der Präimaginalstadien der Pilzmücken (Diptera). II. Eidonomie der Larven. — *Zool., Jb., Anatomie,* **101**: 217-392.
Priesner, H. (1968): Thysanoptera. — *Handb. Zool.,* **4** (2), 5: 1-32.
Rähle, W. (1970): Untersuchungen an Kopf und Prothorax von Embia ramburi Rimsky-Korsakow, 1906 (Embioptera). — *Zool. Jb., Anatomie,* **87**: 248-330.
Rousset, A. (1966): Morphologie cephalique des larves de Planipennes (Neuroptera). — *Mém. Mus. natn. Hist. nat., Paris.,* **42**: 1-199.
Schaller, F. (1970): Collembola. — *Handb. Zool.,* **4** (2), 12: 1-72.
Shvanvich, B. N. — Шванвич, Б. Н. (1949): *Курс общей энтомологии.* — Изд. "Советская Наука", Москва–Ленинград, 900 пп.
Sjöstedt, Y. (1900-1926): Revision der Termites Afrikas. 1-2. — *K. svenska vetensk-Akad. Handl.,* **34** (4): 1-236; **38** (4): 1-120.
Snodgrass, R. E. (1935): *Principles of insect morphology.* — McGraw-Hill Book Company, New York, 667 pp.
Steinmann, H. (1964): Szitakötö lárvák — Larvae Odonatorum. — In: *Magyarország Állatvilága (Fauna Hungariae),* V (7). Akadémiai Kiadó, Budapest, 48 pp.
Steinmann, H. (1967): Tevenyakú fátyolkák, Vízifátyolkák, Recésszárnyúak és Csörös rovarok — Raphidioptera, Megaloptera, Neuroptera és Mecoptera. — In: *Magyarország Állatvilága (Fauna Hungariae),* XIII (14). Akadémiai Kiadó, Budapest, 204 pp.
Steinmann, H. (1968): Álkérészek — Plecoptera. — In: *Magyarország Állatvilága (Fauna Hungariae),* V (8). Akadémiai Kiadó, Budapest, 185 pp.

Steinmann, H. (1970): Tegzesek — Trichoptera. — In: *Magyarország Állatvilága (Fauna Hungariae)*, XV (19). Akadémiai Kiadó, Budapest, 400 pp.

Steinmann, H. and Zombori, L. (1981): *An atlas of insect morphology.* — Akadémiai Kiadó, Budapest, 248 pp.

Štusák, J. M. (1957): A contribution to the knowledge of some last nymphal instars of the Czechoslovakian lace bugs (Hemiptera — Heteroptera, Tingidae). — *Acta Soc. ent. Cechosl.*, **54** (2): 132–141.

Stys, P. and Soldan, T. (1980): Retention of tracheal gills in adult Ephemeroptera and others insects. — *Acta Univ. Carol., Biologica* (1978): 409–435.

Vásárhelyi, T. (1982): Alianates dudichi sp. n. from the Island St. Lucia (Heteroptera). — *Acta zool. hung.*, **28:** 157–170.

Wéber, H. (1966): *Grundriss der Insektenkunde.* — VEB Gustav Fischer Verlag, Jena, 428 pp.

Weidner, H. (1970): Zoraptera. — *Handb. Zool.*, **4** (2), 13: 1–12.

Weidner, H. (1970): Isoptera. — *Handb. Zool.*, **4** (2), 13: 1–147.

Weidner, H. (1972): Copeognatha (Psocodea). — *Handb. Zool.*, **4** (2), 18: 1–94.

Wenk, P. (1953): Der Kopf von Ctenocephalus canis (Curt.) (Aphaniptera). — *Zool. Jb., Anatomie,* **73:** 103–164.

Westwood, J. O. (1837): Characters of Embia, a genus of instects allied to the White Ants (Termites). — *Trans. Linn. Soc. Lond., Zoology,* **17:** 369–374.

Williams, M. and Kosztarab, M. (1970): Morphology and systematic study of the first instar nymphs of the genus Lecanodiaspis (Homoptera: Coccoidea). — *Res. Div. Bull. Virginia,* **52:** 1–95.

Wundt, H. (1961): Der Kopf der Larve vom Osmylus chrysops L. (Neuroptera, Planipennia). — *Zool. Jb., Anatomie,* **79:** 557–662.

Yang, S. P. and Kosztarab, M. (1967): A morphological and taxonomical study on the immature stages of Antonina and of the related genera (Homoptera: Coccoidea). — *Res. Div. Bull. Virginia,* **3:** 1–73.

Yuasa, H. (1923): A classification of the larvae of the Tenthredinoidea (Hymenoptera). — *Contr. ent. Lab. Univ. Illinois.* **69:** 1–140.